Current Topics in Chirality - From Chemistry to Biology

Edited by Takashiro Akitsu

Published in London, United Kingdom

IntechOpen

Supporting open minds since 2005

Current Topics in Chirality - From Chemistry to Biology
http://dx.doi.org/10.5772/intechopen.92523
Edited by Takashiro Akitsu

Contributors
Ishfaq Ahmad Rather, Rashid Ali, Ngoc-Van Thi Nguyen, Kyeong Ho Kim, Kim-Ngan Huynh Nguyen, Kien TRung Nguyen, Hassan Y. Aboul-Enein, Beata Zagórska-Marek, Shikha Agarwal, Ayushi Sethiya, Nusrat Sahiba, Jindra Valentova, Lucia Lintnerová, Susana Porcel-García, Hirotaka Ihara, Yutaka Kuwahara, Makoto Takafuji, Takashiro Akitsu

First published in London, United Kingdom, 2021 by IntechOpen
IntechOpen is the global imprint of INTECHOPEN LIMITED, registered in England and Wales, registration number: 11086078, 5 Princes Gate Court, London, SW7 2QJ, United Kingdom
Printed in Croatia

British Library Cataloguing-in-Publication Data
A catalogue record for this book is available from the British Library

Additional hard and PDF copies can be obtained from orders@intechopen.com

Current Topics in Chirality - From Chemistry to Biology
Edited by Takashiro Akitsu
p. cm.
Print ISBN 978-1-83968-953-6
Online ISBN 978-1-83968-954-3
eBook (PDF) ISBN 978-1-83968-955-0

For EU product safety concerns:
IN TECH d.o.o., Prolaz Marije Krucifikse Kozulić 3, 51000 Rijeka, Croatia, info@intechopen.com or visit our website at intechopen.com.

Meet the editor

Takashiro Akitsu, Ph.D., is now a professor in the Department of Chemistry, Faculty of Science Division II, Tokyo University of Science, Japan. Studying crystal and electronic structures of chiral copper complexes, he graduated from Osaka University and obtained his Ph.D. in Physical and Inorganic Chemistry in 2000. Dr. Akitsu studied at the Institute for Protein Research (metalloproteins), Keio University (photo and magnetic functional organic/inorganic hybrid compounds), and Stanford University (physical and bioinorganic chemistry) before moving to Tokyo University of Science. He has published 220 articles and book chapters. He has also served as an editorial board member and peer reviewer for many journals and was involved in the organizing committees for several international conferences.

Contents

Preface

In a previous book about chirality, entitled Chirality from Molecular Electronic States published in 2019 [1], I wrote in the Preface:

In chemistry, biology, and physics, "chirality" is an important concept in nature. In chemistry, not only classical stereochemistry but also asymmetric organic synthesis, supramolecular chemistry, construction of bio-related molecules, and molecular recognition have become indispensable structural chemical keywords. However, in view of synthetic chemistry and its structural chemistry, chemistry dealing with chirality in relation to the more fundamental electronic state is still a minority. This book focuses on chiroptical spectroscopy, structural or physical features, and theoretical computation of chirality.

In contrast to the previous book, this new book about chirality includes contributions from authors in many fields of natural science, providing a wider overview. The book's focus is chirality and organic chemistry, including synthesis and reactions. For

Figure 1.
Mirror Image formula on a piece of "memory bread," which is a famous Japanese cartoon character's item to memorize letters by transferring from the notebook and eating (for examination of mathematics).

example, the chapter by Ali et al. describes anion-π catalysts in asymmetric synthesis, which is an area of supramolecular chemistry. So far, some important reactions that produce chiral centers have taken place on the π-acidic surface of the anion-π catalyst. This clarifies the importance of anion-π catalysis in the area of asymmetric synthesis.

As I write in the introductory chapter, "Chirality is a concept that is related not only to organic chemistry but also to each field of natural science in a narrow sense, and it is considered that awareness of hierarchy is important for universal and comprehensive understanding." In this way, I would like to continue "fixed-point observations" of chemistry and various sciences, which are three-dimensional mirror images (**Figure 1**), if there is an opportunity to do so in the future..

Takashiro Akitsu
Department of Chemistry,
Faculty of Science,
Tokyo University of Science,
Tokyo, Japan

References

[1] Akitsu T (Ed.), Chirality from Molecular Electronic States, InTech (Rijeka, Croatia) (2019). 978-1-78985-156-4. (Introductory Chapter: Chirality from Molecular Electronic States. pp1-7)

Chapter 1

Introductory Chapter: Chirality

Takashiro Akitsu

1. Introduction

Chirality is a concept that is related not only to organic chemistry but also to each field of natural science in a narrow sense, and it is considered that awareness of hierarchy is important for universal and comprehensive understanding. For example, on a small scale, from the decay of elementary particles, the angular momentum of atoms and light or gravity, organic molecules, biochemical or organic-chemical reactions, crystal structures, supramolecular composite structures, and living organisms, and even galaxy vortices should be considered in this context. Here are some dreams (hypothesis) and excellent examples of interdisciplinary application of "chirality".

2. Angular momentum arising chirality in molecular level

A few years ago, in order to overcome such hierarchy, it was merely a "thought experiment" and a "working hypothesis," but I applied the force of a classical magnetic field to a molecule on a quantum mechanical scale. Like "molecular machines" attracting recent attention, mechanical laws and mechanical properties at the level of molecules and aggregates that are originally below the nanometer scale are used as if they were parts on the centimeter scale, or they are excluded. It can be said that chemistry and material science, which are controlled by the field, are new fields that can be expected to develop in the future. Expanding from a quantum chemistry point of view to a slightly macroscopic point of view, if a molecule is treated as "a kind of rigid body with a certain size and shape" in the framework of classical mechanics, "torque" and "rotational inertia" are among the mechanical properties. Focusing on "momentum", it is expected to discuss the transmission of rotation and the interaction with the orbital angular momentum, which are useful for nanotechnology.

Apart from the control of electronic states and molecular structures by the external field in the conventional quantum chemistry framework, polarized ultraviolet light and magnetism are used as means of external field control, and metal complex molecules, metal complexes-high realization of control of droplets (microdroplets) in molecular/protein composite materials and microfluidics containing composite materials, and mechanical properties such as "torque" and "rotational inertia momentum", which are indicators of the control, are used in molecular design. The goal of this "dream" is to establish a new concept of rotational motion with chirality to be introduced as a new parameter.

There is a magnetic field effect on red blood cells as a phenomenon that has been known for a long time in a similar substance system [1]. Hemoglobin, in which heme iron is paramagnetic and the protein site is diamagnetic, receives resistance from the fluid as it flows through the blood vessels, but when a specific magnetic field acts, blood flow is promoted. This is because not only the paramagnetic part where a direct

IntechOpen

magnetic force is expected, but also the diamagnetic part that occupies most of it undergoes an orientation change that lowers the energy of the magnetic moment due to the repulsion from the magnetic field, making it easier to move (**Figure 1**).

At this time, if the erythrocytes themselves, which are nanoscale molecular aggregates, are regarded as rigid bodies having a constant mass, size, and flat shape (momentum of inertia),

1. external magnetic force

2. magnetic momentum

3. Lorentz force

4. rotational momentum of inertia

5. rotational torque

6. fluid resistance reduction

In order to design a material that is easy to rotate (prone to generate the same torque), the classical mechanical "momentum", how to incorporate the external field response site in the molecule, and the mass-distance distribution from the center of gravity to the end of the molecule, it should be necessary to introduce new molecular parameters (not found in conventional quantum chemistry, *etc.*).

For the optimized structure of the complex after photoisomerization, "the sum of the products of the atomic weight of each atom and the distance between the center of gravity of the molecule (which may be defined as the center of paramagnetic metal)" will be calculated. This is a quantity corresponding to the rotational moment of inertia of classical mechanics, and is newly introduced as a parameter on the molecular side in order to discuss the correlation with the quantity corresponding to the torque realized by the change in external field orientation.

Ferrofluids have long been known as functional materials related to the dynamics of fluids and continuums. However, it was expanded from the fact that the magnetic material is a uniformly dispersed droplet, starting from the (single molecule) "rotational moment of inertia" and "rotational torque", which are important parts of this research. It is significant in proposing parameters for molecules that are completely different from the concept of discussing the propagation of effects to composite materials and droplets in microspace.

Further, as a precedent example in which a rotational motion is caused by light, there is a circularly polarized laser irradiation on metal nanoparticles, metal nanorods, etc., but the size of the target substance composed of metal atoms is too large. Physical quantities such as angular momentum and torque may be able to challenge the limits of size and hierarchy reflected in the chirality of molecules and aggregates.

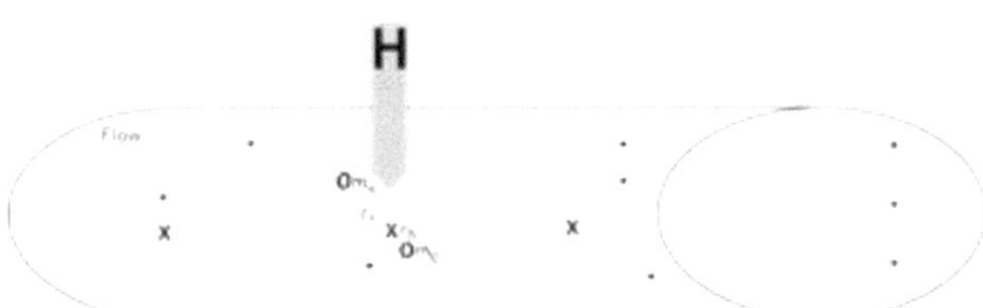

Figure 1.
*Schematic representation of the concept of hemoglobin in blood flow (arrows) under magnetic field (**H**).*

Circularly polarized light with spin angular momentum results in a spiral orientation with chirality. In recent years, with "optical vortex" having orbital angular momentum, dynamics such as torque and light pressure when a spiral structure is transferred to a circular region centered on a singular point on the material surface. Then I noticed that the current situation is that the chemical understanding of "treating molecules classically" may be inadequate. In addition, when attempting to grow a single crystal of a complex-encapsulating protein in a droplet, the "force acting on an object" including gravity is referred to by referring to the protein crystal growth and molecular orientation under zero gravity or strong magnetic field conditions. I came up with the idea that the analogy to molecules and nano-aggregates would be effective.

Unfortunately, verification research of this hypothesis has not yet been conducted to date.

3. Chiral-induced spin selectivity (CISS) effect for hybrid materials

By the way, as a chirality phenomenon that can still cross the hierarchy, "chiral-induced spin selectivity (CISS)" proposed by Professor Ron Naaman is one of the most notable concepts at present [2]. The information gained through the experiment will be meaningful to the science, though this description is merely a "proposal" level.

As mentioned below, injecting electrons into oxidoreductase through oligopeptide which works as a spin filter is the first ever experiment, and it is crucial to cooperate with Dr. Naaman for this research. They believe several years of experience on metal complexes and profound knowledge about the CISS Dr. Naaman has can lead to a promising outcome. One of results they already expect to have is that if they detect the change of chiral property affected by the subtle transformation of metal complex and oligopeptide, they can observe and analyze the electrochemical change.

Their cooperative research covers the fields of the surface physics and the biochemistry, and their results of fixing the electron transfer pathway between electrode and enzyme, and of control of magnetism that metal complex has by the external magnetic field, would be applied to other fields. Also, the CISS effect will offer new possibilities for a better understanding of process of spin selection in the biology and spin electronics applications.

The injection of electrons into laccase using the CISS effect is the first trial of its kind. If this experiment was conducted successfully, the results would contribute to other fields including the spin electronics.

Then they were planning to apply the CISS effect to metal complex to magnetize spins from outside to control them while performing electrochemical measurement. The CISS effect is that in chiral molecules only spins which are orientated in a certain direction can be transferred smoothly. This effect has been studied a lot in recent years with Dr. Naaman, as the leading researcher. With Dr. Naaman's cooperation, who is the leading researcher of this effect, it is particularly noteworthy that difference in chirality shows different spin polarization, which suggests that this effect is applicable to a system where laccase is used (**Figure 2**).

Prepare a measuring device described, then inject spin polarized electrons into laccase where metal complex is present. Place oligopeptide and the metal complex between the electrode and the laccase. They use the oligopeptide and the metal complex, which let spin polarized electrons be transferred smoothly, so that the increase of amperage should be expected. They also analyzed how spin polarized electrons affect the way electron-transfer to the enzyme is occurred. As

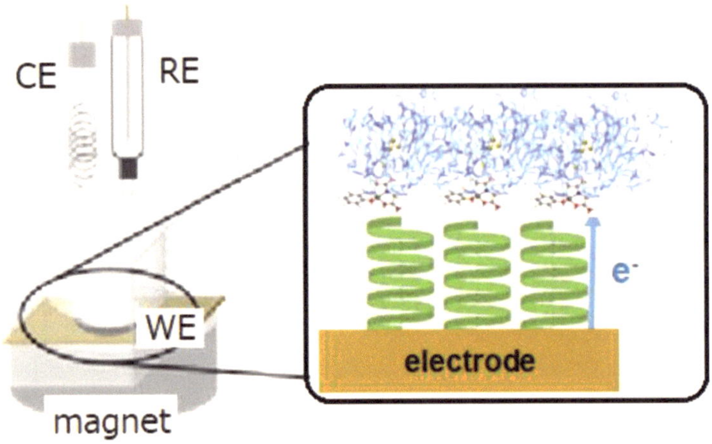

Figure 2.
Proposed of measuring system for CISS effect of redox proteins (laccase).

a preliminary experiment, they conducted electrochemical measurement using the above method without the metal complex. The result shows that the chirality of peptide affects the electron-transfer to the laccase. Following this outcome, they expect the follows. If they insert the metal complex between the oligopeptide and laccase, with the metal complex being fixed inside the hydrophobic pockets of laccase, the metal complex would be fastened firmly to the laccase. Then the electron-transfer to the Cu-T1, which proceeds oxygen reduction reaction, can be carried out selectively. In addition to that, they will examine how the electron-transfer would change when creating an external magnetic field, and they will also analyze it by comparing the use of ferromagnetic or photoresponsive metal complexes.

Some reports have been submitted regarding the injection of electron spins only into membrane protein or oligopeptide in those days, but there has not been any report on the injection of the electron spins into oxidoreductase. The results from the preliminary experiments suggest there is a possibility of realizing the electron-transfer without being influenced by the external magnetic field. Their idea has an intension of examining the electron-transfer in depths from a new perspective by injecting the electron spins into laccase using the CISS effect to elucidate the difference electrochemically between other mediate complexes. The analysis about behavior of the electrons will contribute to the development of biochemistry and a field of spintronics device.

Unfortunately, actual research using laccase only has not yet been published to date.

4. Concluding remarks

In this way, "chirality" may be found in many fields of natural sciences and technology potentially. The main purpose of this plan about the CISS effect study was just to improve electron transfer between the electrode and the enzyme in enzymatic biofuel cell using the CISS effect. Along with that, they are also planning to reveal from a spin-polarized point of view how magnetic field and chiral property affect the electron transfer reactions. Also, if we achieve the implementation of the CISS effect, the result would be applied "chirality" for other areas such as energy materials science and medical biochemistry.

Author details

Takashiro Akitsu
Department of Chemistry, Faculty of Science, Tokyo University of Science, Tokyo, Japan

*Address all correspondence to: akitsu2@rs.tus.ac.jp

References

[1] Higashi T, Takeuchi T: Oyo Butsuri 1998; 67, 1142-1145. https://doi.org/10.11470/oubutsu1932.67.1142 and references therein.

[2] Michaeli K, Kantor-Uriel N, Naaman R, Waldeck DH, Chem. Soc. Rev. 2016; 45, 6478 – 6487. https://doi.org/10.1039/C6CS00369A and references therein.

Chapter 2

Low Melting Mixture of L-(+)-Tartaric Acid and *N,N′*-Dimethyl Urea: A New Arrival in the Green Organic Synthesis

Rashid Ali

Abstract

After the first report of deep eutectic mixtures by the team of Abbott in 2003, the advent of green synthesis has been progressively changing the way synthetic chemistry is thought and also taught. Since then, a plethora of efforts worldwide have been taken to stretch the ideas of sustainable as well as environmentally benign approaches to do the crucial synthetic organic transformations under operationally simple yet effective conditions. Although, till date, several green synthetic strategies for examples ultrasound, microwaves, flow as well as grindstone chemistry etc., and green reaction media (e.g. ionic liquid, water, $scCO_2$, and so forth) have successfully been invented. But a low melting mixture of L-(+)-tartaric acid (TA) and *N,N′*-dimethylurea (DMU), usually plays a double and/or triple role (solvent, catalyst, and/or reagent), though still infancy but enjoys several eye-catching properties like biodegradability, recyclability, non-toxicity, good thermal stability, tunable physiochemical properties, low vapor pressure as well as reasonable prices in addition to the easy preparation with wide functional groups tolerance. To this context, keeping the importance of this novel low melting mixture in mind, we intended to reveal the advancements taken place in this wonderful area of research since its first report by the Köenig's group in 2011 to till date. In this particular chapter, firstly we would disclose the importance of the green synthesis followed by a brief description of deep-eutectic solvents (DESs) particularly emphasizing on the role of L-(+)-TA and DMU from modern synthetic chemistry perspective.

Keywords: low melting mixture, 1,3-dimethyl urea, tartaric acid, sustainable chemistry, synthesis

1. Introduction

As can be inspected from the literature, there were rising concerns in the mid-1980s, regarding the plentiful of the waste being produced by the chemical industry [1, 2]. A paradigm change was undoubtedly desirable, from the old-fashioned perceptions of reaction selectivity, and efficiency which emphasis fundamentally on the chemical yields, to one that allocates the value to the enlargement of the bulk raw materials exploitation, avoidance of the utilization of the hazardous chemicals/reagents/solvents and also preventation of the waste being formed within the

IntechOpen

boundaries of environmental awareness. To this context, in 1987, the term sustainable development was coined by Brundtland, in his report; he mainly focused on the emergence of the societal and industrial development to afford an escalating global population with a suitable value of life in such a way that it should be sustainable over a long period of time [3]. Therefore, complete balance necessities to be found among the three Ps-planet, people, profit i.e. environmental impact, societal equity and economic development. More specifically, in sharp contrast to the green chemistry, sustainable development also comprises an economic factor and if a technology is not economically viable, it could not be sustainable for a long time. Remarkably, a tremendous curiosity in sustainable and green progress, united with a cultivating concern for the climate change, has engrossed attention on resource competence and also driving the shift from a conventional linear flow of bulk materials in a "take–make–use–dispose"economy, towards the greener and even more sustainable globular economy. Interestingly, since the 12 principles of green chemistry (*Prevention of waste; Atom economy; Less hazardous chemical syntheses; Designing safer chemicals with fewer hazards; Safer solvents and auxiliaries; Design for energy efficiency; Use of renewable feedstocks; Reduce derivatives during synthesis; Catalysis; Design for degradation; Real-time analysis for pollution prevention; Inherently safer chemistry for accident prevention*), postulated by Anastas & Warner in 1998 [4], scientists around the world are trying to reduce the volatile organic solvents (VOCs) which generally are the major portion (approx. 80% of the total content) of the reaction vessel as compare to the reactants/reagents, and also has the tendency to escape into the environment, which in turns contribute to ozone depletion as well as smog in urban areas, and hence extremely dangerous for mankind [5]. Therefore, great efforts are being put forward to reduce these hazardous VOCs, and the corrosive acid catalysts, participating in the reaction to make the chemical processes even more sustainable and environmentally friendlier [6]. To this context, over the past few decades, several surrogates for instance water, ionic liquids, supercritical fluids, and switchable solvents in addition to many green strategies such as ultrasound, flow chemistry, biocatalysis, microwaves, and multi-component etc., have successfully been developed [7–9]. Generally, water is thought to be an archetype solvent as it enjoy many classical properties, nonetheless it not only suffer from insolubility issues with the majority of organic compounds but also has a difficulty of removing it after the completion of the reaction because of its high boiling point, and even in many cases compounds gets decompose into the water in addition, some reactions for example amidations and transesterifications, can not be performed in water because of competing product hydrolysis [10]. On the other hand, supercritical fluids which possess low vapor pressure along with the advantages of easy disposal/removal, and recycling, are thought to be the best eco-friendly substitutes of VOCs, but, they requires more sophisticated equipment to perform the reaction. To this context, researchers turn their attention towards the ionic liquids due to their remarkable physiochemical properties but owing to their high cost as well as involvement of the non-renewable resources besides purification before their usage make them of bit doubt from green perception [11]. Consequently, bearing in mind, the urgency of the suitable alternative green solvents in place of conventional solvents to carry out the crucial synthetic transformations for sustainable development in R and D and also for the chemical industry, Abbott's 2003 discovery of the deep eutectic solvents (DESs), also known as low transition temperature mixtures (LTTMs), or low-melting mixtures (LMMs) or deep eutectic ionic liquids (DEILs), has become one of the strongest pillars to the modern synthetic community. Generally, in DES, two/three components are mixed in an appropriate amount to generate a eutectic mixture with lower melting point as compare to the individual components being used [12–15]. As a consequence, an infinite number of melts involving different compositions/components with distinctive properties like price of the raw materials, melting point, polarity, dissolving ability etc., can be accomplished effortlessly. Interestingly, because of the involvement of non-covalent interactions including hydrogen bonds, it has been noticed that the melting points of the DESs are generally below 100°C, even some of them are liquid at room temperature, and they have been the role model among the greener solvents over the past two decades to both academic as well as industrial community because of their remarkable properties and benefits such as biodegradability, low cost and low vapor pressure in addition to non-toxicity and good thermal stability. Among the DESs, a low melting mixture of DMU/TA can be regarded as the solvent of the 21st century, as it hold the following features: (i) Generally, it does not require tedious work-up after the reaction is being completed, rather, just filtration after addition of the water to the reaction mixture while hot, furnishes the analytically pure compounds and most of the time no need of chromatographic purification but simple recrystallization provides the pure form of the required products; (ii) the melting mixture can willingly be recovered and recycled several times without any substantial loss in the activity; (iii) the reaction cleanly underwent towards the product formation at faster rate as compare to the known procedure, and mostly better yields are obtained under operationally simple reaction conditions; (iv) No additional catalyst and solvents are needed in this method, as in conventional procedure, generally both, the corrosive catalysts as well as hazardous, flammable, and volatile organic solvents are being employed; (v) No inert atmosphere is required for a reaction to be successfully completed in parallel yields; (vi) This method also provide good selectivity and also exhibit excellent functional group tolerance; (vii) Easy preparation of the melt from the bulk renewable resources and no further purification before its utilization is needed; (viii) improved safety and very simple handling as comparison to the conventional practices.

Bearing all the above mentioned applications and peculiar physiochemical properties of the DMU/TA melt in mind, which we still feel is immature, although employed for a variety of successful reactions for instance Diels-Alder reaction,

Figure 1.
12 principles of green chemistry and the achievements made in DMU/TA.

Stille, Sonogashira, Suzuki, and Heck coupling reactions, Biginelli reaction, 1,3-dipolar reaction, in addition to its applicability for the synthesis of quinolines, arylhomophthalimides, prymidopyrimidinediones, tetrahydropyrimidinones, hydantoins, dihydropyrimidinones, quinazolines, and a variety of functionalized indole systems with excellent selectivity in decent yields. Interestingly, the beauty of this method is its double and triple role in the reaction vessel to facilitate the accomplishment of the reactions in a clean and smooth fashion without the involvement of any catalyst/additives or solvent. In short, after a brief introduction related to the sustainability and green synthetic approaches, herewith, we have tried to display a deep survey of what has already been done in this field, and open the opportunities to the young researches to find out the new advances by employing this DES and also medium engineering might be utilized to optimize the synthetic utility of various other combinations of the DESs. Green chemistry 12 principles as well as the achievements made by employing a low melting mixture of DMU/TA in the domain of synthetic organic chemistry are displayed in the **Figure 1**.

2. Construction of indole systems under a low melting mixture of DMU/TA

The name indole was originated from portmanteau, a combination of both the words, indigo and oleum which was first isolated from the indigo dye, while treating it with oleum [16, 17]. As can be inspected from the literature, indole scaffold, a notable privileged lead bicyclic aromatic system (10π-electrons), formally known as benzopyrrole, have immeasurable potential applications ranging from the broad-spectrum biological (e.g. *anti*-HIV, antiviral, antimicrobial, antidiabetic, antimalarial, *anti*-cholinesterase, anticancer, *anti*-inflammatory, antioxidant, *anti*-tubercular, *anti*-hypertensive, *anti*-convulsant, *anti*-analgesic, and *anti*-depressant activities etc.), agrochemical and clinical applications to the novel therapeutic agents in addition to their usage as dyes, and smart functional materials as well [18–20]. Interestingly, this venerable heterocyclic moiety is not only a part of several important drug molecules and remarkable receptors in host-guest chemistry but also reside in a variety of medicinally active natural products for instance strychnine, reserpine, alstonine etc.; widespread in diverse species of animals, plants, marine organisms, and the part of lysergic acid diethylamide (LSD) as well [21]. More interestingly, they have inimitable property of mimicking the structure of peptides and nicely bind to the enzymes, in addition to exhibit the momentous pharmacological, physiological, synthetic and industrial applications [22]. A list of some important biologically active molecules (**1–12**) containing the indole moiety is depicted in the **Figure 2** [23, 24].

The typical Fischer indolization (FI) reaction involving arylhydrazine (**13**) and aldehyde/ketone (**14**) in the presence of appropriate acid or acid catalyst along with its systematic mechanistic pathway is displayed in the **Figure 3**. Although, a number of pathways were anticipated for the FI, but the one proposed in 1924 by G.M. Robinson and R. Robinson was the most accepted by the scientific community as it was established by both kinetic as well as the spectroscopic means (**Figure 3**) [25]. The mechanism for this particular reaction commence with the activation of the carbonyl carbon of **14** through the protonation with acid/acid catalyst, employed in the operation, which on further reaction with **13** provide the *N*-arylhydrazone intermediate (**17**). Next, the intermediate (**17**) afforded the ene-hydrazine intermediate (**18**) by means of tautomerization, which upon subsequent [3,3]-sigmatropic rearrangement, distracting the aromaticity of aryl ring

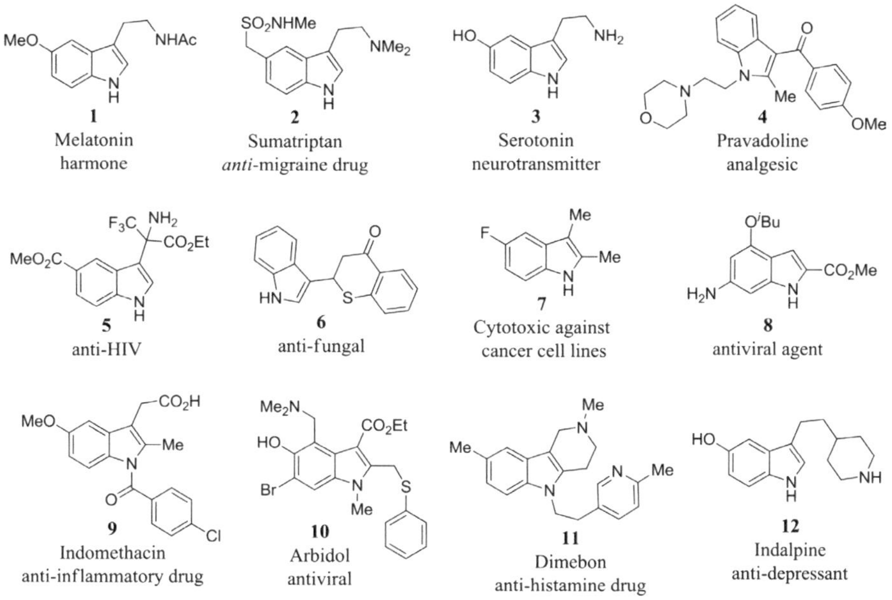

Figure 2.
Structures of some important biologically active indole derivatives.

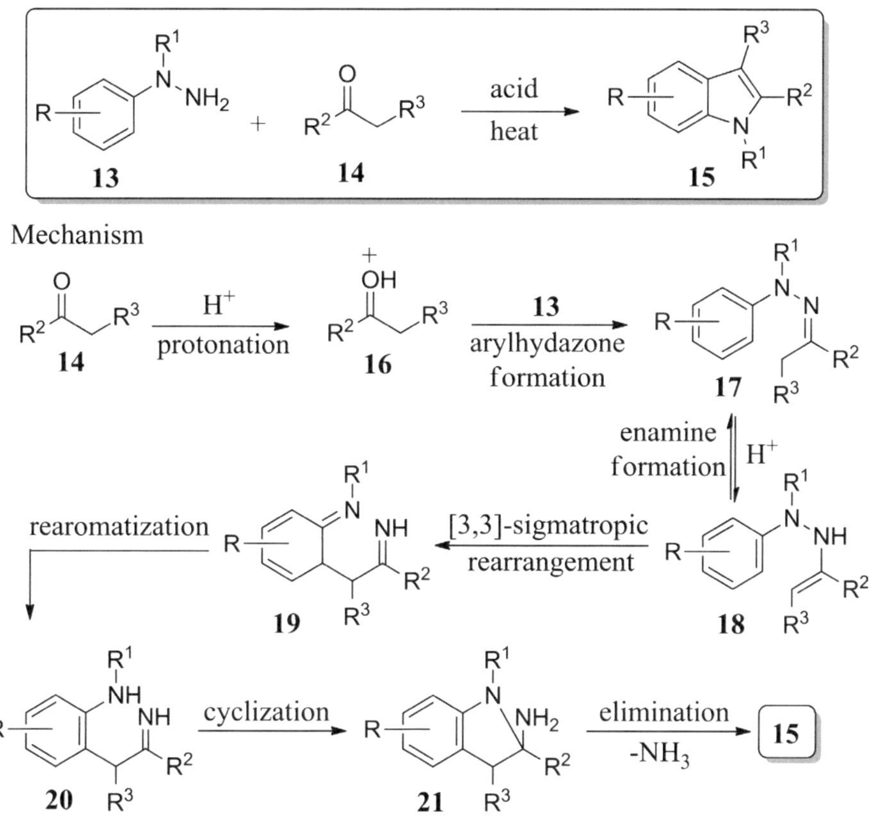

Figure 3.
Representation of indole formation along with the plausible mechanism.

system, followed by rearomatization deliver another intermediate (**20**) through the *bis*-iminobenzyl ketone (**19**). Latter furnishes the required indole derivatives (**15**) by virtue of cyclization followed by the loss of ammonia molecule *via* **21** (**Figure 3**). Interestingly, it has been observed that the reaction conditions as well as the nature of the substrate decide the rate determining step (rds). Generally, ene-hydrazine intermediate (**18**) formation or the [3,3]-sigmatropic rearrangement step has been noticed as the rate-limiting step depending on the situation, as discussed further below. The [3,3]-sigmatropic rearrangement has been observed as rate determining step, in a specific case of α-N-acyl hydrazones in addition to weak acidic solutions as well as when ammonia elimination is prevented due to steric effects [25]. Whileas, in most of the cases including the strong acidic condition favors the ene-hydrazine (**18**) formation as the rds-step of the reaction. More specifically, unsymmetrical 1,l-diarylhydrazines under strong acidic condition provide the indolization at most activated ring (i.e. most susceptible to the protonation), whileas under neutral reaction conditions almost equal amount of isomers are generally being formed.

Accordingly, synthetic chemists have long sought approaches for the construction of indole architectures, and a plethora of methods continue to be reported in this trend [26]. Hardly surprising, to date, a myriad of methods involving both intra- and intermolecular transformations for the construction of indole derivatives, particularly the usage of named reactions such as, Gassman, Bartoli, Thyagarajan, Julia, Schmid, Wender, Couture, Kihara, Nenitzescu, Engler, Saegusa, Liebeskind, Sundberg, Hemetsberger, Magnus, Feldman, Reissert, Makosza, Leimgruber–Batcho, Watanabe, Larock, Yamanaka–Sakamoto, Hegedus–Mori–Heck, Fürstner, Castro, Natsume, Nordlander, and so on, have successfully been employed [27]. But, to our best knowledge, despite its numerous complications, rearrangements, and also mechanistic mysteries, Fischer indole protocol, an old yet effective procedure which involve a pericyclic tool namely, [3,3]-sigmatropic rearrangement, remains the epitome for the scientific community around the globe to assemble diverse indole and its congeners [28]. Although, a variety of acid catalysts for example HCl, AcOH, PPA, $TiCl_4$, $ZnCl_2$, $SOCl_2$, PCl_3, TsOH, H_2SO_4, mont-morilloniteclay zeolite etc., have been employed to synthesize the indole framework using FI protocol, but simple, and eco-friendly methods which involve non-hazardous, inexpensive and easily accessible chemicals as well as reagents utilizing the environmentally benign practices are always of particular interest. In this regard, König's group in 2012, first time reported a green approach by employing the FI strategy under a low melting mixture of dimethyl urea (DMU):L-(+)-tartaric acid (TA) in (7:3) ratio to yield a range of indole derivatives in good-to-excellent yields [24]. The beauty of this particular green method relies on the fact that, a clean low melting mixture is generated just by heating the two components in appropriate amount at much lower temperature than its individual components, and can be used without further purification. Herewith, the low melting mixture, acts as mild acidic catalyst (pH 3.7) as well as solvent to furnish the required indoles with great functional group compatibility and selectivity. As can be seen from an inspection of the **Figure 4**, these authors prepared a range of functionalized indole systems (**22–47**) in decent yields using acyclic and cyclicketones in addition to cyclic enol ethers for instance dihydrofuran and dihydropyran. Fascinatingly, optically active ketone deliver the indole with retention of the configuration. Moreover, indolenines (**31**), was also prepared through this powerful technique in respectable yields under mild reaction conditions (**Figure 4**). Besides, hormone melatonin (**25**) and dimebon (**26**) were also obtained by utilizing this wonderful green approach as a crucial step (**Figure 4**). Inspiring form this simple yet powerful procedure and also from the applications of the indole moiety containing molecules, two years later to this report, in 2014, Kotha and his teammates have successfully employed this strategy for the synthesis of C_2-and C_s-symmetric *bis*-indole

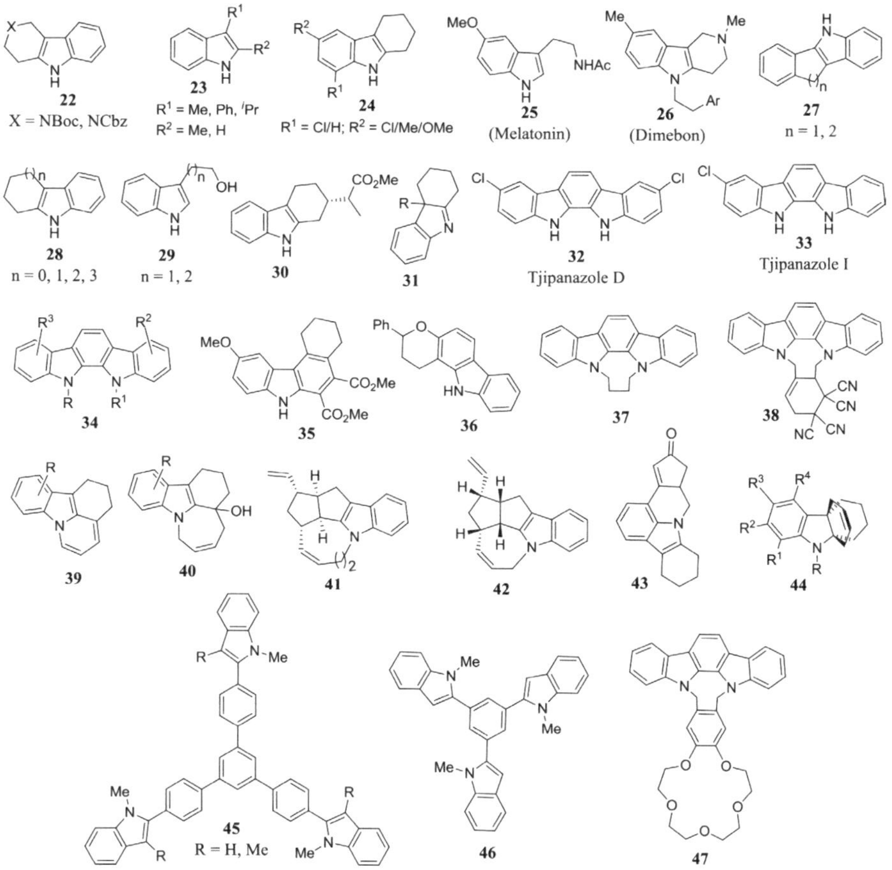

Figure 4.
Indole derivatives constructed using DMU/TA mediated green protocol.

systems (**52, 53, 58, 60–62**) from bicyclo-3,7-diones and 1-methyl-1-phenylhydrazine under DMU/TA (7:3) reaction conditions (**Figure 5**) [29]. Later on, Kotha's team nicely expanded this delightful method for the generation of a variety of carbazole derivatives (**32–35**) including pyrano-carbazole (**36**) and *aza*-cyclophane based carbazoles (**37** and **38**) as depicted **Figure 4** [30–32] in **Figure 4**. Interestingly, utilizing this tactic, they have also prepared carbazole-based natural products such as tijapinazole D (**32**) and tijapinazole I (**33**) in addition to the crown-based indolocarbazole (**47**). Moreover, in the laboratory of Kotha's group, diverse hetero-polycyclic compounds (**39–43**) in addition to the propellane derivatives (**44**) have been assembled by using ring-closing metathesis (RCM) and Fischer indolization in a low melting mixture of DMU/TA as crucial steps, (**Figure 4**) [33–35]. Keeping the importance of C_3-symmetric molecules in medicinal and bioorganic chemistry besides their vital role in material science and technology, the same group has also prepared star-shaped C_3-symmetric compounds **45** and **46** involving cyclotrimerization and DMU/TA mediated indolization approach (**Figure 4**) [36]. Furthermore, as can be inspected from the **Figure 5**, they design and constructed varied cyclophane derivatives (**48, 49, 54, 55** and **59**) through the involvement of the Grignard reaction, RCM and a low melting mixture mediated indolization sequences in respectable yields because of their applicability in supramolecular chemistry [37–41]. In addition to these, Kotha's group has also prepared diverse polycyclic mono- (**50, 56, 63**) and *bis*-indole derivatives (**51, 57, 58, 64**) by means of a deep eutectic mixture of DMU/TA (70:30) under operationally simple reaction conditions [42–45].

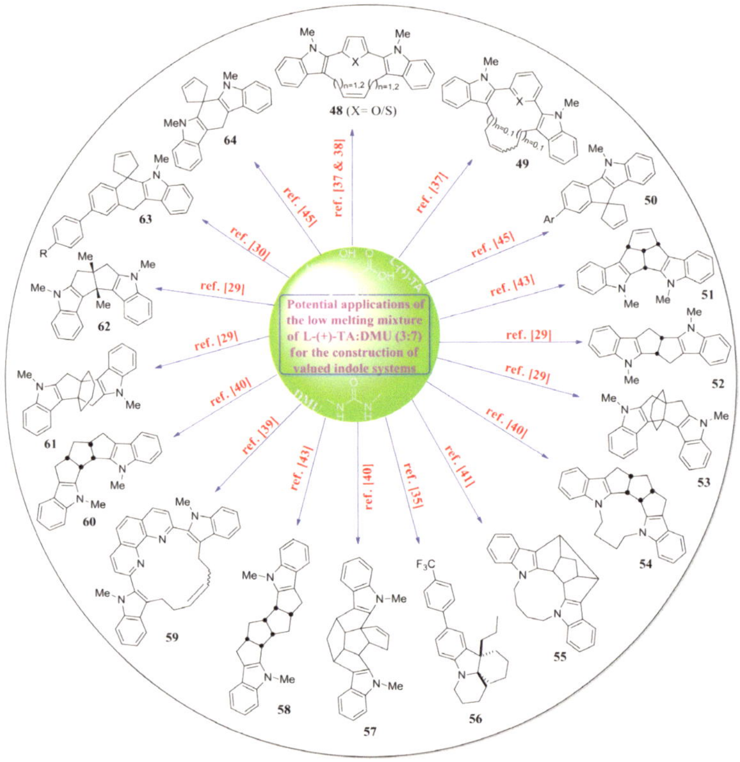

Figure 5.
Diverse indole derivatives assembled via *FI utilizing the DMU/TA mixture.*

3. Synthesis of *bis*-(indolyl)methane, indenylindoles and 2,2-disubstituted indolin-3-ones

In recent past, *bis*-(indolyl)methanes (BIMs) have attracted a tremendous attention of the research community due to their potential applications both in pharmaceuticals and agrochemicals besides their activity in breast tumor cells, bladder cancer and also inhibits proliferation practice as well. Moreover, they display antitumorgenic, antibiotic, antimicrobial activity and *anti*-inflammatory activities etc., and are mostly found in marine natural sources. Fascinatingly, getting inspired from the above applications and also other, if any, the group of Nagarajan constructed the diverse BIMs (**67**) including the natural products arsindoline A and B *via* a green protocol in the presence of DMU:TA (70:30) (**Figure 6**) [46]. Surprisingly, the BIMs were not formed when instead of aldehydes (**66**); cyclicketones (**68**) were treated with indole derivatives in the Kotha's laboratory, rather they received indenylindoles (dienes) **69** under parallel reaction conditions (**Figure 6**) [30]. On the other hand, it has been found that, numerous medicinally active natural and non-natural products possess 2,2-disubstituted indolin-3-one scaffold in addition to their usage as the key building blocks in the total synthesis of diverse indole alkaloids. In this regard, Xie's group involved a deep eutectic mixture of DMU/TA to furnish a range of 2,2-disubstituted indolin-3-one derivatives (**72**) as displayed in the **Figure 6** [47].

Synthesis of heterocyclic compounds has always been of prime importance to the research community because of their vital role in a numerous areas ranging from material sciences and technology to the pharmaceutical and agrochemical

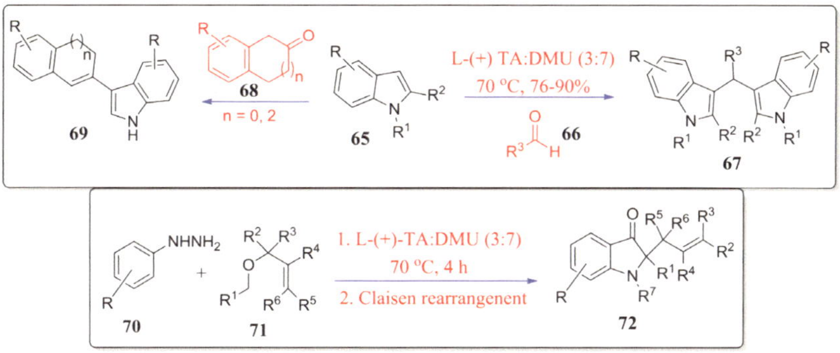

Figure 6.
Synthesis of bis-*(indolyl)methane, indenylindoles and 2,2-disubstituted indolin-3-ones utilizing DMU/TA mediated green approach.*

industries. To the best of our knowledge, to date, a choice of drugs containing heterocyclic scaffolds are available in the world market, and many hundred are under clinical trials around the globe. Therefore, there are always high demands to develop novel strategies for the generation of heterocyclic systems particularly involving milder reaction conditions in an environmentally friendlier manner from easily assessable bulk materials. To this context, although a number of methods having several advantages and disadvantages are available in the literature but in recent years, the deep eutectic solvents have changed the scenario of modern synthetic chemistry by providing a plethora of green approaches towards the construction of these valuable molecules. Among the heterocyclic systems, quinoline scaffold has received a considerable amount of interest because of its availability in a plethora of bioactive molecules. A very simple yet effective green procedure for the synthesis of a variety of quinoline derivatives (**75**) have been developed by Zhang and his co-workers with the involvement of a low melting mixture of DMU:TA (70:30) in moderate-to-excellent yields in a Friedländer fashion (**Figure** 7) [48]. On the other hand, the Biginelli procedure, a multi-component reaction, has been employed for assembling the dihydropyrimidinones (DHPMs) under a green reaction conditions by Köenig's team because of their utility in calcium channel blockers and also as HIV inhibitors and anticancer agent (**Figure** 7) [49]. Captivatingly, this procedure works equally well with masked aldehydes to furnish the required DHPMs in reasonable yields. In another study, the same group has utilized this powerful green methodology for assembling diverse functionalized pyrimidopyrimidinedione derivatives (**85**) with the help of Biginelli reaction in which the low melting mixture play a triple role such as solvent/catalyst/reagent (**Figure** 7) [50]. In this study, although, they have tried several low melting mixtures but DMU/TA in a ratio of 7:3 provided the best results.

In a separate study, Krishnakumar *et al.,* has reported a green chemical procedure for the construction of *N*-arylhomophthalimides (**83**) by employing the Michael addition reaction of the Michael-donor (homophthalimides) **82** with Michael-acceptor (chalcones) **81** in DMU:TA low melting mixture (**Figure** 7) [51]. In this report, the authors have screened various reaction conditions but the mention conditions provided the good results for both electron withdrawing as well as electron donating groups containing contestants.

The hydantoin and its congeners are the key scaffolds from biological point of view as they are the part of various molecules which exhibit a range of activities for instance antidepressants, antiulcer, antidiabetic agents, anticonvulsant, antiarrhythmic, and antiviral etc. Moreover, this moiety also play a significant role in agrochemistry, cosmetic industry, dye-sensitized solar cells, chiral auxiliaries and also used

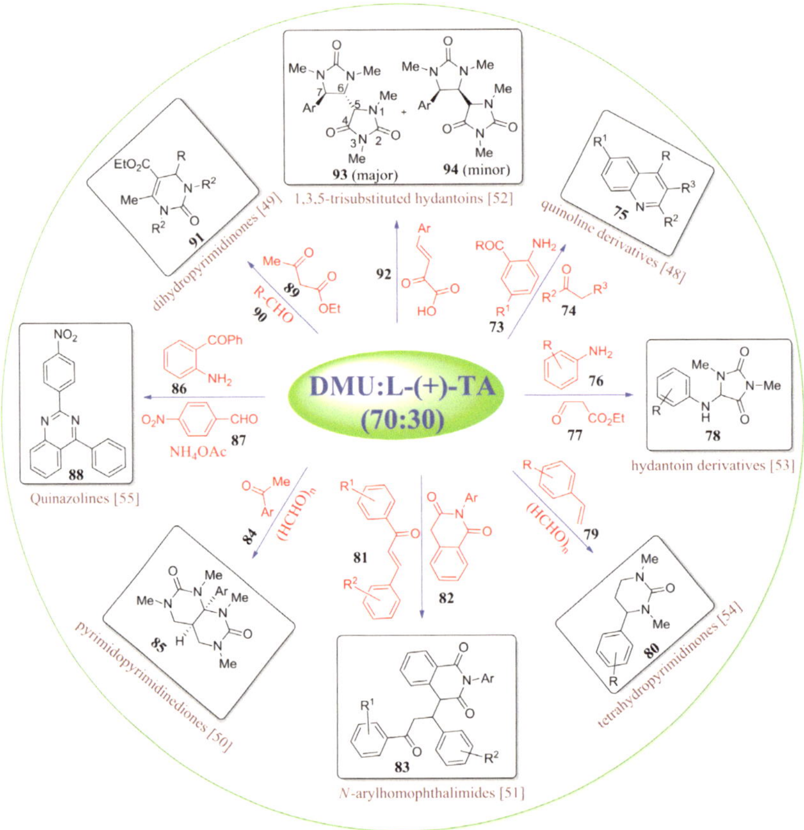

Figure 7.
A variety of heterocycles assembled by employing the DMU/TA melt.

as the intermediates for the generation of enantiomerically pure natural and non-natural α-amino acids by means of the dynamic kinetic resolution. Therefore, keeping the consequence of these molecules in mind, König's group in 2013 developed a simple and eco-friendly method for the synthesis of 1,3,5-trisubstituted hydantoin derivatives (**93/94**) in excellent yields by means of DMU/TA melt-mediated green approach (**Figure 7**) [52]. Interestingly, during their experimentation, they noted down good diastereoselectivity in which *anti*-isomers were isolated in major amount whileas *syn*-diastereomers were obtained as minor products, confirmed by nuclear overhauser effect (NOE) and X-ray analysis means. On another front, quite recently, Kotha's team has reported mono-hydantoins as well as thiohydantoins by means of three component reaction under low melting mixture of DMU/TA with electron neutral, electron donating, and electron withdrawing groups possessing aniline derivatives (**Figure 7**) [53]. Finally, the tetrahydropyrimidinones (**80**) and quinazoline derivatives (**88**) have been reported, by the groups of Baskaran and Zhang, respectively by employing the same low melting mixture of DMU/TA under similar reaction conditions as can be inspected from the **Figure 7** [54, 55].

4. Conclusions and outlook

In summary, a novel method involving DMU/TA as a low melting mixture has comprehensively been revealed in this chapter, depicting its pivotal role in the heart of modern synthetic organic chemistry particularly for the generation of a variety of valuable heterocyclic systems. Herein, we have disclosed, a decade advancements made in this field since its inspection (2011). As discussed above in detail, this simple, environmentally benign, cost effective, and productive method has already been shown its impact in the domain of modern preparative chemistry in general, and green chemistry in particular. We assure that this chapter based on greener transformations, will not only help the readers for complete understanding of a low melting mixture of DMU/TA, and its contribution towards the vital synthetic organic transformations, but also would inspire the motivated researchers to exploit the masked opportunities. More importantly, this method might provide a new way to the chiral catalyst mediated reaction since herewith, chiral tartaric acid is part of the melt, and may act as a valuable handle for the generation of chirality in a molecule under the operation.

Acknowledgements

Dr. Rashid Ali is grateful to DST-SERB New Delhi for financial support (Project File no. ECR/2017/000821). In addition, he also thanks Jamia Millia Islamia, New Delhi, India, for providing the necessary research facilities.

Conflicts of interest

The author declare no conflicts of interest.

Author details

Rashid Ali
Organic and Supramolecular Functional Materials Research Laboratory, Department of Chemistry, Jamia Millia Islamia, Jamia Nagar, Okhla, New Delhi, India

*Address all correspondence to: rali1@jmi.ac.in

References

[1] Sheldon R.A, *Organic synthesis-past, present and future*. *Chem. Ind.* (London) 1992:903.

[2] Sheldon R.A, *Fundamentals of green chemistry: efficiency in reaction design*. *Chem. Soc. Rev.* 2012:41:1437.

[3] *Report of the world commission on environment and development: our common future*; Oxford University Press, Oxford, U.K., 1987.

[4] Anastas P.T, Warner, J.C. *Green chemistry: theory and practice*, Oxford University Press, Oxford, 1998.

[5] Ramón D.J, Guillena G, *Deep eutectic solvents: synthesis, properties, and applications*, Wiley-VCH Verlag GmbH & Co. KGaA, 2019.

[6] Welton T, *Room-temperature ionic liquids. solvents for synthesis and catalysis*. *Chem. Rev*. 1999:99:2071.

[7] Butler R.N, Coyne A.G, Water: nature's reaction enforcer-comparative effects for organic synthesis "in-water" and "on-water" *Chem. Rev*. 2010:110:6302.

[8] Jessop P.G, Ikariya T, Noyori R, Homogeneous catalysis in supercritical fluids *Chem. Rev*. 1999:99:475.

[9] Jutz F, Andanson J.-M, Baiker A, Ionic liquids and dense carbon dioxide: a beneficial biphasic system for catalysis. *Chem. Rev*. 2011:111:322.

[10] Kitanosono T, Masuda K, Xu P, Kobayashi S, Catalytic organic reactions in water toward sustainable society. *Chem. Rev.* 2018:118:679.

[11] Zhang Q, Zhanga S, Deng Y, Recent advances in ionic liquid catalysis. *Green Chem*. 2011:13:2619.

[12] Perna M.P, Vitale P, Capriati V, Deep eutectic solvents and their applications as green solvents. *Curr. Opin. Green Sustain. Chem.* 2020:21:27.

[13] Liu P, Hao J.-W, Mo L.-P, Zhang Z.-H, Recent advances in the application of deep eutectic solvents as sustainable media as well as catalysts in organic reactions. *RSC Adv*. 2015:5:48675.

[14] Smith E.L, Abbott A.P, Ryder K.S, Deep eutectic solvents (DESs) and their applications *Chem. Rev*. 2014:114:11060.

[15] Kotha S, Chakkapalli C, Application of Fischer indolization under green conditions using deep eutectic solvents. *Chem. Rec*. 2017:17:1039.

[16] Fischer E, Jourdan F, Ueber die hydrazine der brenztraubensäure. *Ber. Dtsch. Chem. Ges*. 1883:16:2241.

[17] Robinson B, *The Fischer indole synthesis*, Wiley-Interscience, New York, 1982.

[18] Sundberg R.J, *Indoles*, Academic Press, San Diego, CA, 1996.

[19] Sundberg R.J, '*The chemistry of indoles*', AcademicPress, New York, 1970, Chapter III.

[20] Palmisano G, Penoni A, Sisti M, Tibiletti F, Tollari S, Nicholas K.M, Synthesis of indole derivatives with biological activity by reactions between unsaturated hydrocarbons and aromatic precursors. *Curr. Org. Chem.* 2010:14:2409.

[21] Heravi M.M, Rohani S, Zadsirjan V, Zahedi N, Fischer indole synthesis applied to the total synthesis of natural products. *RSC Adv.* 2017:7:52852.

[22] Kumari A, Singh R.K. Medicinal chemistry of indole derivatives: Current to future therapeutic prospectives. *Bioorg. Chem.* 2019:89:103021.

[23] Colella M, Degennaro, L, Luisi R, Continuous flow synthesis of heterocycles: a recent update on the flow synthesis of indoles. *Molecules*, 2020:25:324.

[24] Gore S, Baskaran S, König B, Fischer indole synthesis in low melting mixtures. *Org. Lett.* 2012:14:4568.

[25] Robinson G.M, Robinson R, The mechanism of E. Fischer's synthesis of indoles. *J. Chem. Soc., Trans.* 1924:**125**:827.

[26] Gribble, G.W, Recent developments in indole ring synthesis-methodology and applications. *Contemp. Org. Synth.* 1994:145.

[27] Gribble G.W, Recent developments in indole ring synthesis-methodology and applications. *J. Chem. Soc., Perkin Trans. 1.* 2000:1045.

[28] Hughes D.L, Progress in the Fischer indole reaction. a review. *Org. Prep. Proced. Int.* 1993:25:607.

[29] Kotha S, Chinnam A.K, Anomalous behaviour of *cis*-bicyclo[3.3.0]octane-3,7-dione and its derivatives during twofold Fischer indole cyclization using low-melting mixtures. *Synthesis,* 2014:301.

[30] Kotha S, Ali R, Saifuddin M, Diversity-oriented approach to natural product inspired pyrano-carbazole derivatives: Strategic utilization of hetero-Diels–Alder reaction, Fischer indolization and the Suzuki–Miyaura cross-coupling reaction. *Tetrahedron,* 2015:71:9003.

[31] Kotha S, Aswar V.R, Chinnam A.K, One-pot synthesis of carbazoles from indoles via a metal free benzannulation. *Tetrahedron Lett*. 2017:58:4360.

[32] Kotha S, Saifuddin M, Aswar V.R. Diversity-oriented approach to indolocarbazoles via Fischer indolization and olefin metathesis: total synthesis of tjipanazole D and I. *Org. Biomol. Chem*. 2016:14:9868.

[33] Kotha S, Ravikumar O, Diversity-oriented approach to carbocycles and heterocycles through ring rearrangement metathesis, Fischer indole cyclization, and Diels–Alder reaction as key steps. *Eur. J. Org. Chem*. 2014:5582.

[34] Kotha S, Aswar V.R, Singhal G, Synthesis of tricyclic units of indole alkaloids: application of Fischer indolization and olefin metathesis. *Tetrahedron,* 2017:73:6436.

[35] Kotha S, Aswar V.R, Ansari S, Selectivity in ring-closing metathesis: synthesis of propellanes and angular aza-tricycles. *Adv. Synth. Catal*. 2019:361:1376.

[36] Kotha S, Todeti S, Das T, Datta A, Synthesis and photophysical properties of C_3-symmetric star-shaped molecules containing heterocycles: a new tactics for multiple Fischer indolization. *ChemistrySelect.* 2018:3:136.

[37] Kotha S, Chinnam, A.K, Shirbhate M.E, Diversity-oriented approach to cyclophanes via Fischer indolization and ring-closing metathesis: substrate-controlled stereochemical outcome in RCM. *J. Org. Chem*. 2015:80:9141.

[38] Kotha S, Chinnam A.K, Shirbhate M.E, Design and synthesis of hybrid cyclophanes containing thiophene and indole units via Grignard reaction, Fischer indolization and ring-closing metathesis as key steps. *Beilstein J. Org. Chem*. 2015:11:1514.

[39] Kotha S, Shirbhate M.E, Chinnam A.K, Sreevani G, Synthesis of phenanthroline and indole based hybrid cyclophane derivatives via ring-closing metathesis. *Heterocycles*, 2016:93:399.

[40] Kotha S, Chinnam A.K, Ali R, Hybrid macrocycle formation and spiro annulation on *cis-syn-cis*-tricyclo[6.3.0.0^{2,6}]undeca-3,11-dione and its congeners via ring-closing metathesis. *Beilstein J. Org. Chem.* 2015:11:1123.

[41] Kotha S, Cheekatla S.R, Chinnam A.K, Jain T, Design and synthesis of polycyclic bisindoles via Fischer indolization and ring-closing metathesis as key steps. *Tetrahedron Lett.*, 2016:57:5605.

[42] Kotha S, Chinnam A.K, Design of aza-polyquinanes via Fischer indole cyclization under green conditions. *Heterocycles*, 2015:90:690.

[43] Kotha S, Chinnam A, K, Sreenivasachary N, Ali R, Design and synthesis of polycyclic indoles under green conditions via Fischer indolization. *Indian J.Chem.* 2016:55B:1107.

[44] Kotha S, Ali R.A, simple approach to *bis*-spirocycles and spiroindole derivatives via green methods such as Fischer indolization, ring-closing metathesis, and Suzuki–Miyaura cross-coupling. *Turk. J. Chem.* 2015:39:1190.

[45] Kotha S, Ali R, Srinivas V, Krishna N.G, Diversity-oriented approach to spirocycles with indole moiety via Fischer indole cyclization, olefin metathesis and Suzuki–Miyaura cross-coupling reactions. *Tetrahedron*, 2015:71:129.

[46] Jella R.R, Nagarajan R, Synthesis of indole alkaloids arsindoline A, arsindoline B and their analogues in low melting mixture. *Tetrahedron*, 2013:69:10249.

[47] Xia Z, Hu J, Gao Y.-Q, Yao, Q, Xie W, Facile access to 2,2-disubstituted indolin-3-ones via a cascade Fischer indolization/Claisen rearrangement reaction. *Chem. Commun.* 2017:53:7485.

[48] Ma F.-P, Cheng G.-T, He Z.-G, Zhang Z.-H, A new and efficient procedure for Friedländer synthesis of quinolines in low melting tartaric acid-urea mixtures. *Aust. J. Chem.* 2012:65:409.

[49] Gore S, Baskaran S, Köenig B, Efficient synthesis of 3,4-dihydropyrimidin-2-ones in low melting tartaric acid–urea mixtures. *Green Chem.* 2011:13:1009.

[50] Gore S, Baskaran S, Köenig B, Synthesis of pyrimidopyrimidinediones in a deep eutectic reaction mixture. *Adv. Synth. Catal.* 2012:354:2368.

[51] Krishnakumar V, Vindhya N.G, Mandal B.K, Khan F.-R.N, Green chemical approach: low-melting mixture as a green solvent for efficient michael addition of homophthalimides with chalcones. *Ind. Eng. Chem. Res.* 2014:53:10814.

[52] Gore S, Chinthapally K, Baskaran S, König B, Synthesis of substituted hydantoins in low melting mixtures. *Chem. Commun.* 2013:49:5052.

[53] Kotha S, Gupta N.K, Aswar V.R, Multicomponent approach to hydantoins and thiohydantoins involving a deep eutectic solvent. *Chem.–Asian J.* 2019:14:3188.

[54] Devi P, Lambu M.R, Baskaran S, A novel one-pot method for the stereoselective synthesis of tetrahydropyrimidinones in a low melting mixture. *Org. Biomol. Chem.* 2020:18:4164.

[55] Zhang Z.-H, Zhang X.-N, Mo L.-P, Li Y.-X, Ma F.-P. Catalyst-free synthesis of quinazoline derivatives using low melting sugar–urea–salt mixture as a solvent, *Green Chem.* 2012:14:1502.

Chapter 3

Gold Catalyzed Asymmetric Transformations

Susana Porcel García

Abstract

In this chapter, the strategies developed to attain asymmetric reactions with gold are disclosed. Because of its preferred linear arrangement, to induce asymmetry, gold(I) needs to fulfill one of the following requirements: a) the use of bulky chiral ligands, that create a chiral pocket around the active site, b) the coordination to bifunctional ligands capable to establish secondary interactions with substrates, or c) tight ion pairing with chiral counteranions. On the other hand, gold(III) profits of a square-planar coordination mode, which approaches chiral ligands to substrates. However, its tendency to be reduced leads to difficulties for its applications in catalytic asymmetric transformations. Pioneering works using cyclometaled structures, have found the balance between stability and activity, showing its potential in asymmetric transformations.

Keywords: gold, catalysis, bulky ligands, bifunctional ligands, ion pairing

1. Introduction

Gold was long considered to be unsuitable for catalysis due to bulk gold present a high reluctance to react. Nonetheless, in the last part of the 20th century pioneering studies from different groups, showed that gold in oxidation states I and III has a big potential as catalysts specially, in reactions dealing with the activation of C-C multiple bonds [1–7]. Because gold(III) is prone to be reduced, the majority of gold catalyzed transformations described so far involves gold(I) complexes. It has been evidenced that gold(I) is able to trigger the building of complex molecular frameworks in a few steps, under soft conditions and with a high degree of functional group tolerance [8–15]. Its special properties lay on relativistic effects, that contract its 6 s and 6p orbitals and expands the 5d shell, lowering the energy level of the LUMO which in turn is traduced in a high Lewis acidity [16]. The development of unsymmetric transformations with gold(I), was hampered at the beginning because of its preferred linear coordination mode [17, 18], that keeps away the substrate being modified from the chiral ligand environment. Hopefully have been uncovered successful strategies to circumvent this problem, such us the use of chiral counter anions, or the development of suitable chiral ligand incorporating secondary interactions with substrates which achieve high asymmetric level. Conversely gold(III) has an square planar coordination mode, ideal for approaching close together the substrate being transformed and the asymmetric ligand, however as note before, its tendency to be reduced has restricted its applications. A few examples has appeared very recently arriving at a compromise between reactivity an stability and it is expected to continue growing in next years, as the chemistry of gold(III) continues to be enlarge.

This chapter is an overview of the strategies and ligands employed to achieve chiral transformations with gold. It is organized according to the type of ligands designed [19–22].

2. Gold(I) asymmetric transformations

2.1 Gold(I) asymmetric transformations with diphosphine ligands

The first asymmetric ligands that enabled moderate to good enantiomeric ratios were atropoisomeric bidentate phosphines. The most commons are depicted in **Figure 1**. The importance of these phosphines relay on their relative accessible synthetic procedures and their commercial availability. Along with them, some planar chiral diphosphines, or diphosphines containing asymmetric carbon centers has also been used, although in a minor extent.

One of the reactions more thoroughly studied in gold chemistry is the cycloisomerization of 1,n enynes. Starting from linear pools, this reaction gives access under soft conditions, to otherwise complex synthetic targets. Primary studies over the cycloisomerization of enynes, showed that the alkoxycyclization of 1,6 enynes proceeds with modest values of enantioselectivity using (*S*)-Tol-BINAP as ligand (**Figure 2**, Ec1). It was assumed that the reaction is triggered by a monocationic gold complex, generated *in situ* by halogen abstraction with a silver salt. Upon coordination to the alkyne, the catalyst would promote a 5-*exo-dig* cyclization. As a result, it would be formed a cyclopropyl gold carbene complex, which evolves to the final alkylidene cyclopentene, by nucleophilic attack of methanol to the cyclopropane moiety [23].

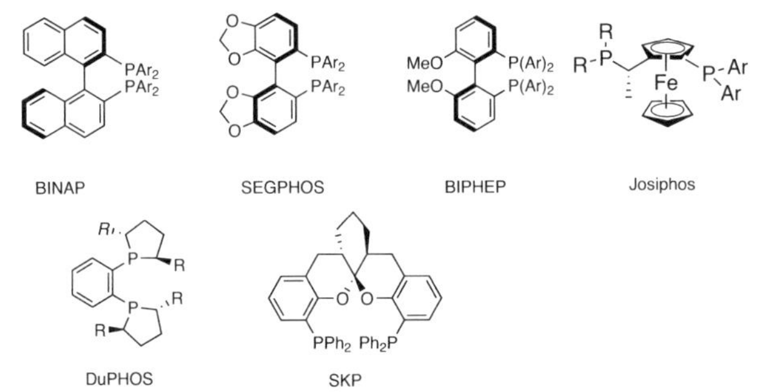

Figure 1.
Common asymmetric biphosphine ligands.

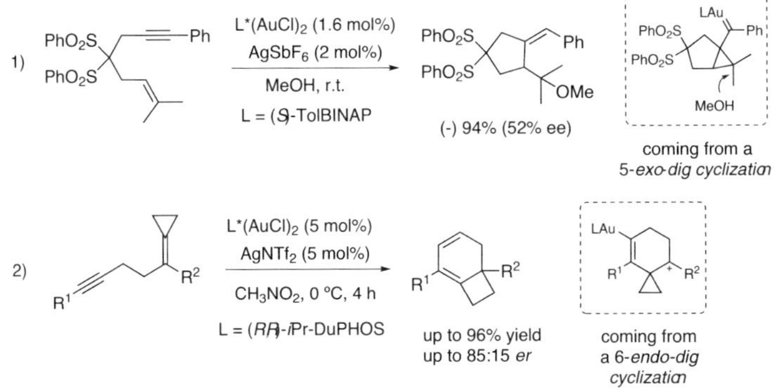

Figure 2.
Asymmetric cycloisomerization of 1,6- and 1,5-enynes.

Improved enantiomeric values were obtained in the cycloisomerization of 1,5-enynes bearing cyclopropyliden moieties (**Figure 2**, Ec2). These substrates led to challenging bicyclo[4.2.0]octanes, by a 6-*endo-dig* cyclization/ring expansion process [24, 25]. It was observed that the enantioselectivity values were importantly affected by the amount of the silver salt employed. The best results were obtained using the complex (*R,R*)-*i*Pr-DuPHOS$(AuCl)_2$ (5 mol%), and AgNTf (5 mol%) as a silver salt. It could be notice that AgNTf itself, is able to catalyze the reaction in some extension, being responsible of the decrease in the enantiomeric ratios. Related to this ring expansion procedure, 1,6-enynes containing a cycloalkoxy unit, have been shown to rearrangement to cyclopentyl-cyclobutanones with high enantioselectivities, when treated with [(*R*)-MeO-DTBM-BIPHEP-$(AuCl)_2$] (3 mol%) and $AgBF_4$ (6 mol%) [26]. On the other hand, 1-allenylcyclopropanols also undergo a cyclization/ring expansion process that affords chiral vinyl cyclobutanones, with good enantiomeric ratios, when treated with (*R*)-MeO-DTBM-BIPHEP$(AuCl)_2$ (2.5 mol%) and NaBARF (5 mol%) as chloride scavenger. The reaction is promoted by Π activation of the allene through gold coordination, and a subsequent Wagner-Meerwein shift [27]. Finally, another strategy for accessing chiral cyclobutanes consists onto an intermolecular [2 + 2] cycloaddition of alkynes and alkenes. This time higher enationomeric ratios were obtained with a Josiphos digold(I) complex (2.5 mol%) and $NaBAr_4^F$ (2.5 mol%). Interestingly, the mechanistic studies carried out in this work, revealed that only one atom of gold is involved in the activation of the alkyne, but the second one is needed to induce enantioselectivity (**Figure 3**) [28].

Chiral cyclopropanes are also amenable with gold complexes by olefin cyclopropanation with diazo compounds (**Figure 4**). Thus, cyclopropanes with vicinal all-carbon quaternary stereocenters can be assembled by reaction of diazooxindoles with α-CH_2F styrenes, using a spiroketaldiphosphine digold(I) complex. This reaction benefits from hydrogen bond interaction with the solvent, particularly fluorobencene forms a strong C-F···H-N interaction, that lower the activation barrier of

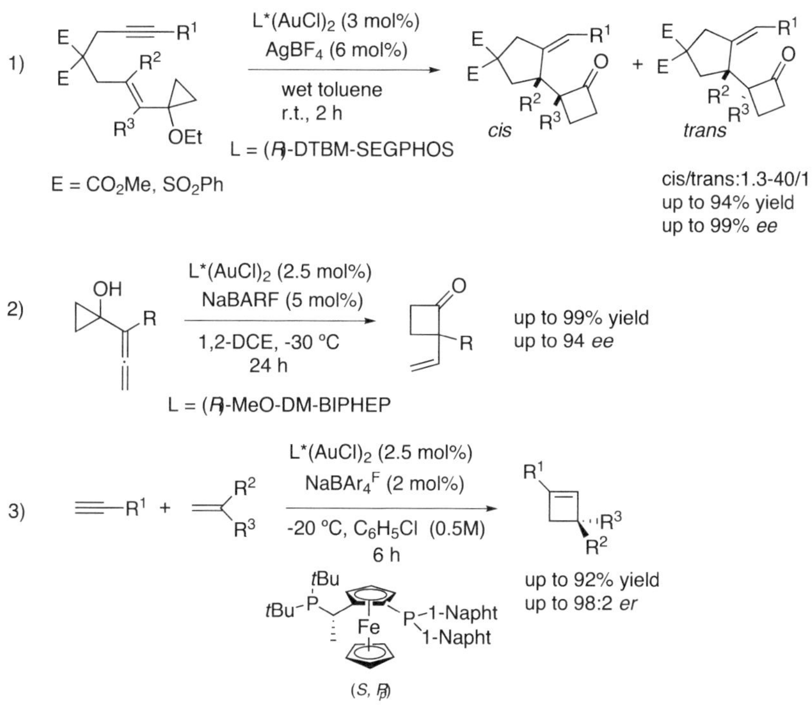

Figure 3.
Asymmetric cyclobutanes synthesis.

the reaction. Yields up to 93% were obtained, with enantioselectivities over 90% and diastereoselectivities higher of 20:1 in all cases [29].

Enantioselective hydroetherification of alkynes is possible by desymmetrization of prochiral phenols containing a P-stereogenic center (**Figure 5**, Ec.1). It has been observed that bisphenols and dialkyne phosphine oxides, undergo a 6-*endo-dig* cyclization with (*S*)-DTBM-SEGPHOS$(AuCl)_2$ complex, leading to chiral cyclic phosphine oxides. The yields of the reaction maintained up to 97% and the enantioselectivites close to 99%. This reaction is an efficient and practical tool to achieve compounds with P-sterogenic centers [30]. Notably, the same complex has been used for the synthesis of planar-chiral ring-fused ferrocenes, starting from *ortho*-alkylnylaryl ferrocenes (**Figure 5**, Ec. 2) [31]. Finally, along with this alkyne activation protocols, (*R*)-DTBM-SEGPHOS$(AuCl)_2$ has been efficiently applied in asymmetric Picted-Spengler reactions between tryptamines and arylaldehydes [32].

2.2 Gold(I) asymmetric transformations with monophosphine ligands

In some reactions catalyzed by chiral digold complexes, better performances were obtained by generation of monocationic instead of dicationic species. This fact points that in those cases the role of the second atom of gold may be just steric, or that it may be involved in secondary interactions with substrates. With this in mind, there has been an increasing interest in developing monophosphine chiral ligands. One of the monophosphines that have exerted better enantioselectivities, are

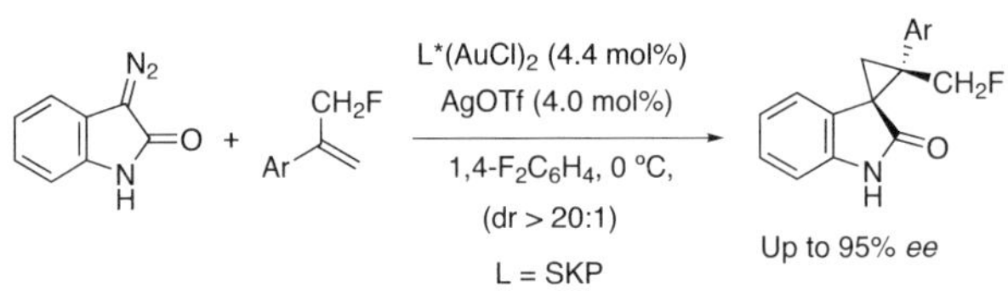

Figure 4.
Asymmetric synthesis of cyclopropanes with diazooxindoles.

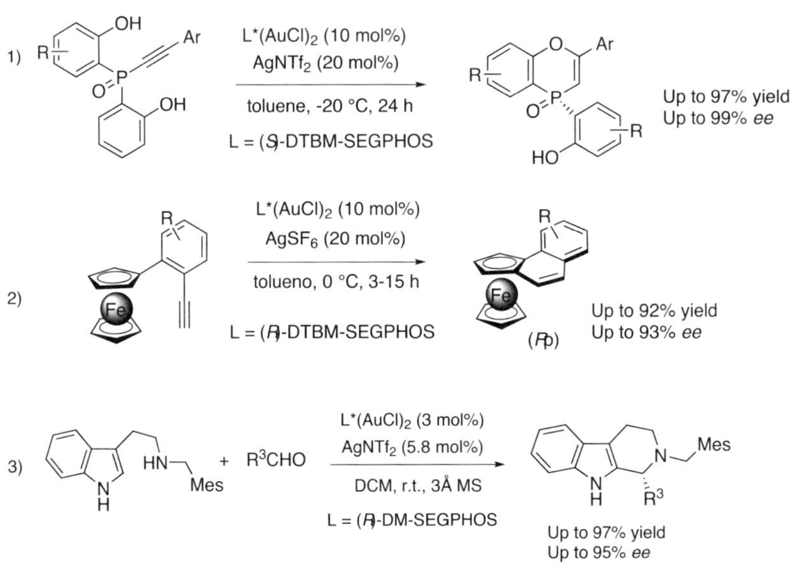

Figure 5.
Asymmetric synthesis of cyclic phosphine oxides, ring-fused planar chiral ferrocenes and tetrahydo-β-carbolines.

monophosphines bearing a chiral sulfinamide moiety that can stablish secondary interactions with the substrates. These ligands offer the advantage of being easily modified, as they can be modularly synthesized. For example, the MING-PHOS family is synthesized by a two-step sequence (**Figure 6**, Ec1), that consists in the condensation of an arylphoshine aldehyde with chiral *tert*-butylsulfinamide, followed by the stereodivergent addition of RLi or RMgX. This way diasteromeric sulfinamide monophosphines can be isolated. MING-PHOS ligands has been applied over a variety of reactions, thus they have shown to catalyze the enantioselective [3 + 3] cycloaddition of 2-(1-alkynyl)-alk-2-en-1-ones with nitrones (**Figure 6**, Ec. 1). The reaction furnished furo[3,4-*d*] [1,2]oxazines, with high diasteroselectivity (> 20:1) and stereoselectivity. Interestingly, both types of enantiomers could be isolated by using a pair of diasterosisomeric MING-PHOS [33, 34]. Replacing nitrones by 3-stylindoles, cyclopenta[c]furans were obtained with similar values of diastereo- and entioselectivities (**Figure 6**, Ec. 3) [35]. A variation of the MING-PHOS family that incorporates adamantyl groups at the phosphorous atom (XIA-PHOS

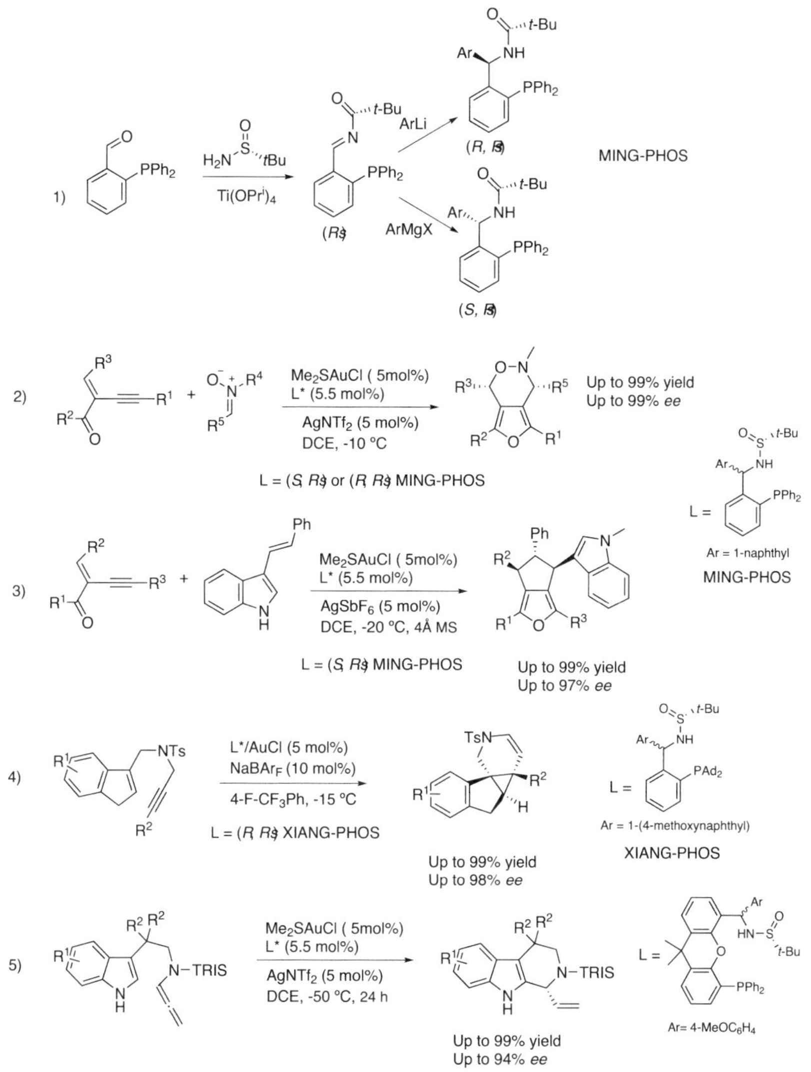

Figure 6.
(1) Synthesis of chiral sulfinamides. (2) Asymmetric reactions with chiral phosphine sulfinamides.

family), has been employed for the synthesis of fused polycycles. Thus, the intramolecular cyclopropanation of indenes or trisubstituted alkenes, led to polycyclic compounds containing two vicinal all-quaternary stereogeneic centers with excellent yields and enantioselectivities (**Figure 6**, Ec. 4) [36]. N-allenamides attached to the indol nuclei could be cyclized to chiral tetrahydrocarbolines, by using PC-PHOS family. This family of ligands combine the well-known Xant-Phos phospine with a chiral sufinamide, affording high levels of entantioselectivity (**Figure 6**, Ec.5).

Along with chiral phosphine sulfinamides, other chiral bifunctional monophosphine ligands have been described. Based on remote cooperative effects, it have been designed axially chiral monophosphines containing a chiral basic center that can stablish secondary interactions with substrates. These types of ligands have been used to obtain asymmetric 2,5-dihydrofurans with excellent values of enantio- and diasteroselectivity, starting from alkynols through isomerization to chiral allenols and subsequent cyclization (**Figure 7**) [37].

Another interesting approach that relays in secondary interactions, consists in the synthesis of phosphines containing a biphenyl scaffold connected to a C_2-chiral pyrrolidine moiety (**Figure 8**). Because of the bulky substituents at the phosporous atom, upon complexation, the P-Au-Cl axis remains parallel to the biphenyl moiety, approaching the gold center to the asymmetric unit. This way it is created a chiral pocket in which the substrate is encapsulated. These ligands have been applied to the cyclization of 1,6-enynes, giving rise to high enantiomeric ratios. DFT calculations showed that the enatioselectivity of the reaction, relays on π-π interactions between the substrate and the ligand. It could be observed opposite enantioselectivities, depending on the position of the aromatic ring in the substrate being cyclized. This chemistry has been applied to the total synthesis of tree members of the carexane family [38].

Finally, phosphahelicenes has also been used to induce asymmetry in the cyclization of 1,6-enynes. These ligands contain a menthyl at phosphorous as the chiral auxiliar. The phosphorous atom racemize at room temperature, and after complexation with a LAuCl precursor, are obtained two epimeric gold complexes; one where the gold atom is disposed toward the helical scaffold (*endo* complex) and another where the gold atom is disposed on the opposite face (*exo* complex). *Endo* isomers give higher enantioselective values since locate closer the metal to the helical moiety (**Figure 9**) [39].

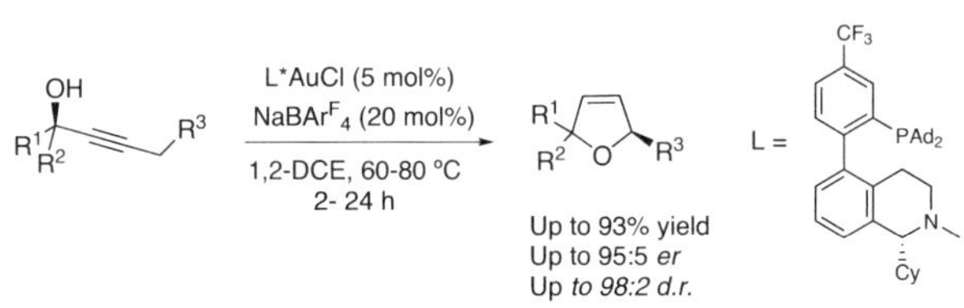

Figure 7.
Asymmetric synthesis of dihydrofuranes with axially chiral bifunctional monophosphines.

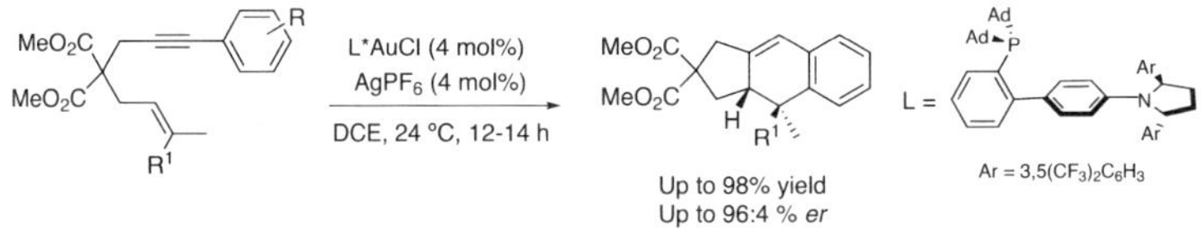

Figure 8.
Asymmetric cyclization of 1,6enynes with biphenyl C_2 chiral pyrrolidine phosphines.

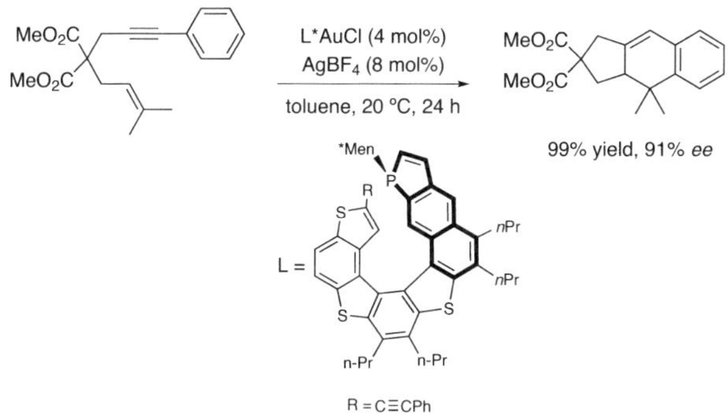

Figure 9.
Asymmetric cyclization of 1,6-enynes with phosphahelicenes.

2.3 Gold(I) asymmetric transformations with phosphoroamidites

Phosphoroamidites are modulable monodentate ligands that exerts good levels of enantiomeric ratios in gold catalysis. The firsts example of asymmetric transformations employing phosphoramidites, where applied to the cyclization of allenes. Using phoshoroamidite ligands based on BINOL scaffold, it was shown that allenedienes undergo a formal (4 + 3) cycloaddition reaction leading to bicyclic compounds via an allylic cation. The carbene derived from this cation, can evolve via a 1,2-H migration shift, affording 5,7-fused bicyclic compounds, or by a ring contraction leading to 6,7-fused bicyclic compounds. The presence of substituents at the end of the allene favors the formation of 6,7-fused bicyclic compounds. The reaction is totally diastereselective and proceeds with high values of enantioselectivity (**Figure 10**, Ec. 1) [40]. Other BINOL derived ligands have been used in the

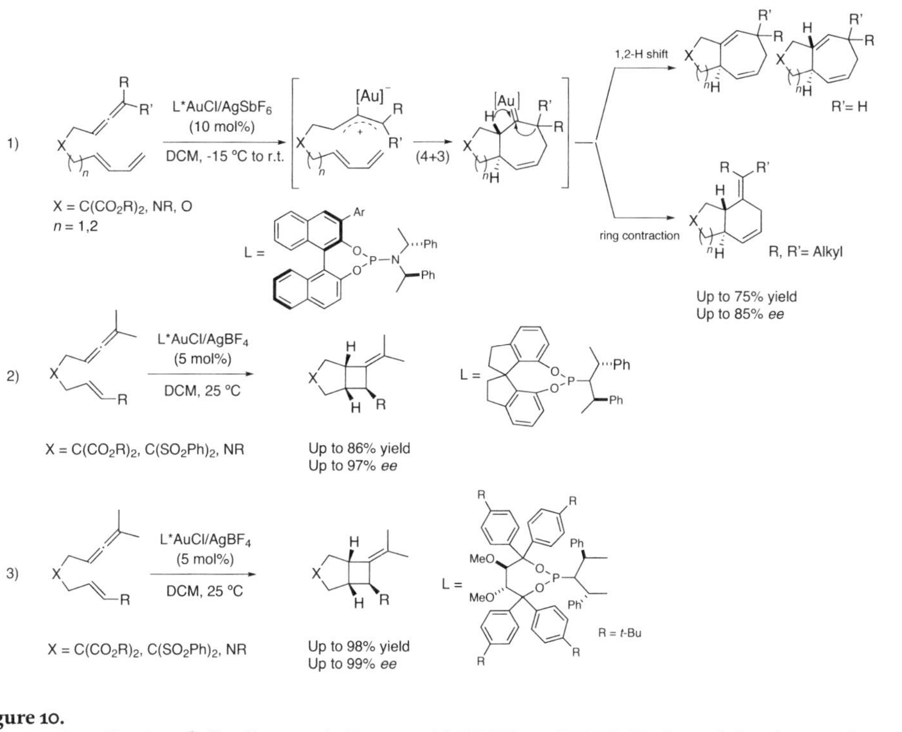

Figure 10.
Asymmetric cyclization of allendienes and allenenes with BINOL and TADDOL-derived phosphoroamidites.

cyclization of allenes. Thus, it has been shown that, allenenes undergo a (2 + 2) cycloaddition reaction furnishing 5 + 4 bicyclic compounds with excellent enantioselective values (**Figure 10**, E. 2) [41]. Along with BINOL, TADDOL-derived phosphoramidites has shown excellent performance in asymmetric reactions catalyzed by gold. This scaffold creates a conic cavity of C_3 symmetry around the gold center. One of the better TADDOL-derived phosphroamidites bears an acyclic backbone. This type of ligands exerts excellent values of enantioselectivity in a variety of gold catalyzed reactions, in particular allenenes undergo a (2 + 2) cycloaddition reaction with excellent levels of asymmetry [42].

After these initial examples, BINOL derived phosphoroamidites have been used in several relevant organic reactions, such as hetero-Diels-Alder reactions (**Figure 11**, Ec. 1), where the chiral gold(I) complex acts as a Lewis acid activating urea-based diazene dienophiles (**Figure 11**, Ec.1) [43], or in the (3 + 2) annulation of 2-(1-alkynyl)-2-alken-1-ones with *N*-allenamides (**Figure 11**, Ec. 2) [44]. This reaction is proposed to procceds by generation of an all-carbon 1,3-dipole and subsequent (3 + 2) annulation at the proximal C=C bond of the alleneamide.

Looking for more electrophilic phosphorous centers, recently TADDOL and BINOL have been used as chiral scaffolds in α-cationic phosphonites. These ligands incorporate an imidazolium, or a related cationic heterocyclic moiety, directly bounded to phosphorous. The cationic group increase the Lewis acidity character of the phosphorous increasing the activity of gold upon complexation. By far, these ligands have been used for the synthesis of helicenes via gold catalyzed alkyne hydroarylation reactions, with excelents levels of enantioselectivity (**Figure 12**) [45, 46].

2.4 Gold(I) asymmetric transformations with carbenes

Although in a minor extent than phosphine and phosphoramidites ligands, both acyclic and cyclic N-heterocyclic carbenes have been used in asymmetric gold catalyzed reactions. Acyclic diaminocarbene ligands with a pendant binaphthyl moiety, induce high enantioselective values in gold catalyzed acetalization/cycloisomerization reactions of *ortho*-alkynylbenzaldehydes (**Figure 13**, Ec. 1) [47]. According to DFT calculations, the wide N-C-N angle of the carbene, approaches the binaphthyl unit to the gold center, facilitating an Au-arene interaction, that creates the chiral environment for the enantio-discrimination. N-heterocyclic carbenes (NHC) have

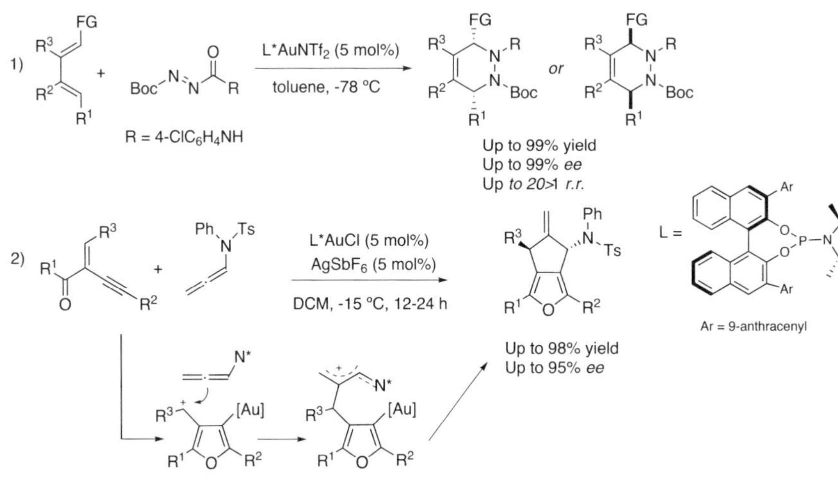

Figure 11.
(4 + 2) and (3 + 2) cyclizations with BINOL-derived phosphoramidites.

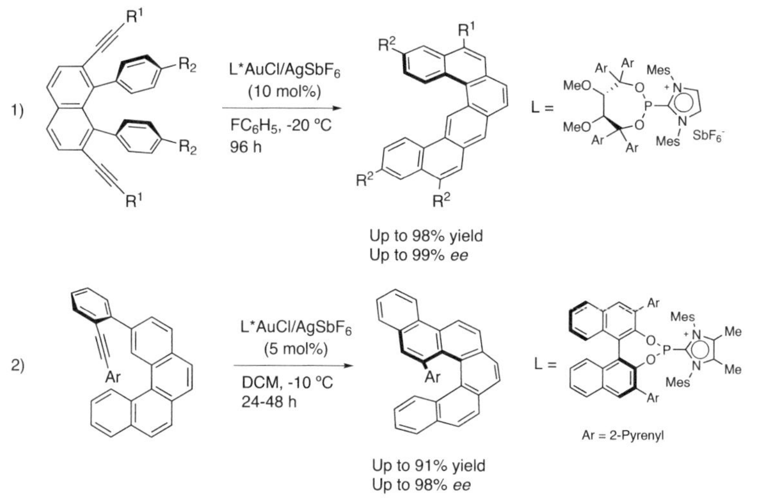

Figure 12.
Asymmetric synthesis of helices with α-cationic phosphonites

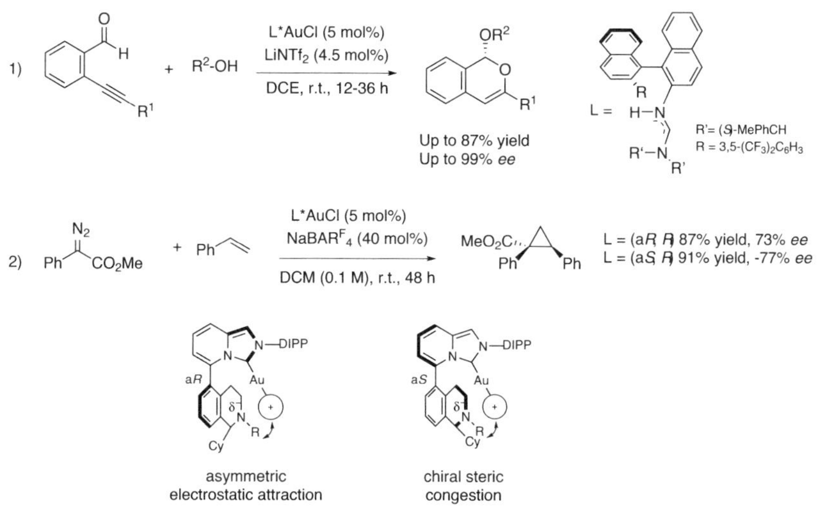

Figure 13.
Asymmetric transformations with carbene ligands.

been used in bifunctional type ligands containing chiral tetrahydroisoquinoline structures. These ligands contain a fluxional biaryl axis, that allow the aryl groups to be orientated orthogonally. After complexation with $AuCl{\cdot}SMe_2$, it was possible to separate two atropoisomer complexes generated, due to the restricted rotation of the biaryl axis in the presence of AuCl. It was observed, that each atropoisomers give rise to opposite enantiomers, as it is illustrated in the cyclopropanation of styrene (**Figure 13**, Ec. 2). In the complex with (a*R*,*R*) configuration, the enantio-discrimination come from an electrostatic attraction effect, between the partially negatively charged ligand nitrogen and the cationic gold center. On the other hand, in the case of the complex with a (a*S*,*R*) configuration, enantio discrimination was attributed to the chiral steric environment posed by the cyclohexyl group [48]. Along with these examples, gold complexes encapsulated in capped cyclodextrin cavities have also shown to catalyze several asymmetric transformations, such us the cycloisomerization of enynes, hydroarylation and lactonizations reactions [49].

2.5 Gold(I) asymmetric transformations with chiral counteranions

The difficulty in creating an asymmetric environment around gold(I), and the cationic nature of gold(I) catalyzed reactions, led to the search of alternatives strategies to induce asymmetry based on ion pairing. Generation of cationic achiral gold complexes, in the presence of chiral counterions, allow inducing asymmetry by transferring the chiral information via formation of tight ion pairs between cationic organogold species and chiral anions. It was first observed, that allenes undergo hydroalkoxylation, hydrocarboxylation and hydroamination reactions with high enantioselective values, using an achiral diphosphine digold complex in the presence of a chiral silver phosphate derived from binaphthol (**Figure 14**, Ec. 1). It was proposed that, the silver phosphate generates a cationic gold(I) complex leaving the chiral phosphate as counteranion, which is responsible for the enantioselectivity observed [50]. The same strategy was applied to the desymmetrization of 1,3-diols (**Figure 14**, EC. 2) [51].

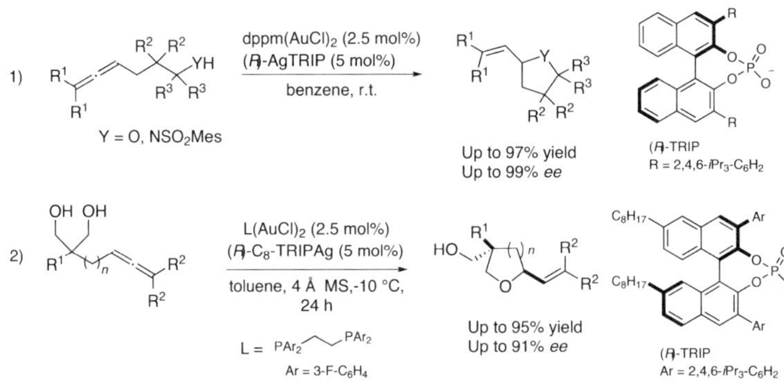

Figure 14.
Asymmetric cyclization of allenes with chiral counterions.

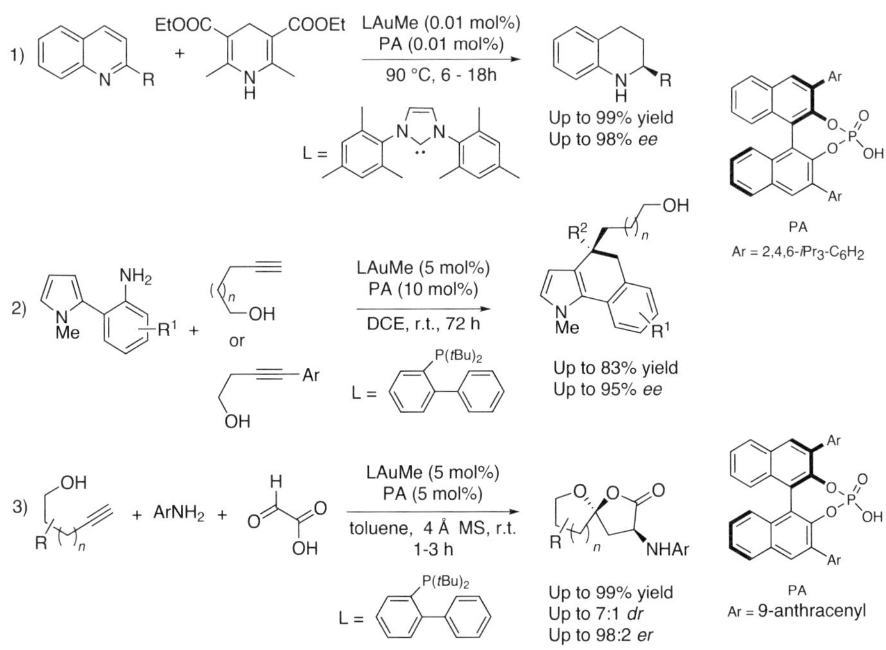

Figure 15.
Asymmetric transformations with chiral phosphoric acids.

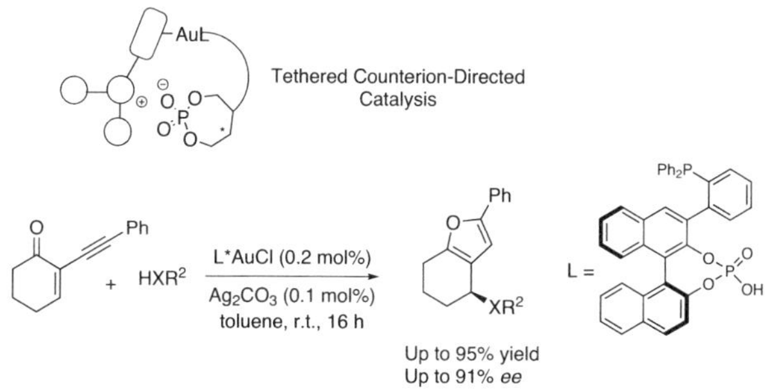

Figure 16.
Asymmetric transformations with phosphoric acid-tethered phosphines.

The cationic gold(I) specie can also be generated with chiral phosphoric acids by protonolysis of complexes precursors with an Au-Me bond. This type of asymmetric induction has been used in enantioselective transfer hydrogenation reactions of quinolines (**Figure 15**, Ec. 1) [52], in the hydroamination-hydroarylation of alkynes (**Figure 15**, Ec. 2) [53] and in the synthesis of spiroacetals among others [54].

In these approximations the degree of enantio-discrimination depends upon the proximity of the counteranion to the cationic gold center. In this sense, recently have been designed new phosphine ligands, thetered to chiral phosphoric acids, with the aim to restring the flexibility of the ion pair. The new phosphoric acid-tethered phosphines have shown excellent levels of enantioselectivity in reactions proceeding through carbocationic intermediates, such us the cyclization-addition of heteronucleophiles to enones (**Figure 16**) [55].

3. Gold(III) asymmetric transformations

Opposite to gold(I), gold(III) complexes have a square-planar geometry that allows ancillary ligands to be closer to the substrate, what made them good candidates for the development of asymmetric transformations. However, its enormous tendency to be reduced, have hampered it use in catalysis. Some recent studies have found the way to stabilize gold(III) centers, while maintaining its catalytic activity, placing them into cyclometalated frameworks. NHC-biphenyl gold(III) complexes with a cyclometalated structure, showed enough stability to catalyze an enantioconvergent kinetic resolution of 1,5-enynes (**Figure 17**, Ec. 1) [56]. In this reaction racemic 1,5-enynes are converted to bicyclo[3.1.0]hexenes with enantiomeric ratios up to 88%. Each enantiomer of the starting 1,5-enyne led to the same bicyclo with different enantioselectivity, making the overall enantioselectivity decrease with the conversion. Because of the latter, the conversions were maintained below, 50%. A related NHC-biphenylene gold(III) catalyst has been applied to enantioselective γ,δ-Diels-Alder reactions. In this occasion enantioselectivities reached 97% and yields were up to 87%. Detailed mechanistic studies revealed that the enantio- discrimination come from non-covalent π-π interactions between the substrate and an aromatic group of the complex (**Figure 17**, Ec. 2) [57].

Other cyclometalated complexes, such as cyclometalated oxazoline gold(III) complexes incorporating a biphenol ligand, have shown to be able to catalyze the asymmetric carboalkoxylation of alkynes. The corresponding 3-alcoxyindanones are obtained with moderate to good enantioselectivities. Remarkably, in this reaction camphorsulonic acid (CSA) activate the gold complex avoiding the need of adding silver salts as activators. Mechanistic studies suggested that the active catalytic specie is formed through protodeauration of one of the oxygens of the biphenol ligand (**Figure 18**) [58].

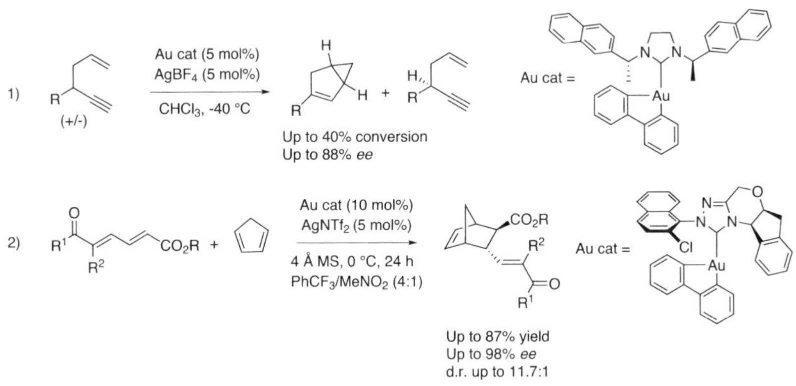

Figure 17.
Asymmetric transformations with chiral NHC-biphenyl gold(III) complexes.

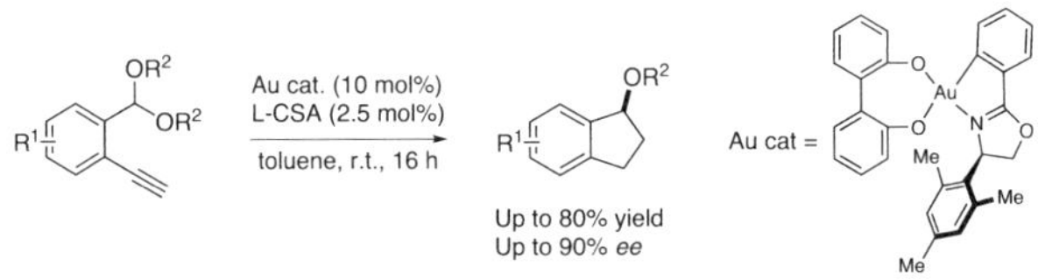

Figure 18.
Asymmetric carboalkoxylation of alkynes with gold(III) complexes.

4. Conclusions

Gold catalyzed asymmetric transformations is an emerging area. Enantioinduction with gold(I) catalysts, is a challenging task due to its preferred linear coordination mode, that place the substrate far from the chiral ancillary ligands. Nonetheless, to date several successful strategies have arose. Some are based on using sterically congested ligands, that create a chiral pocked around the active site, others use bifunctional phosphines, that stablish secondary interactions with the substrate, and finally others are based on using tight ion pairing with chiral anions. On the other hand, asymmetric catalysis with gold(III) is just beginning to be developed. When introducing ancillary ligands around gold(III), a fair balance between stability and activity must be reached. By far, cyclometalated complexes of gold(III) have shown that it is possible to undergo catalytic asymmetric reactions with gold(III), manifesting its great potential. A deeper development is expected in near future with gold(III).

Acknowledgements

This work was supported by CONACyT (A1-S-7805) and DGAPA (IN202017).

Conflict of interest

The authors declare no conflict of interest.

Appendices and nomenclature

Tol-BINAP	2,2'-bis(di-*p*-tolylphenylphosphino)-1,1'-binaphthyl
*i*Pr-DuPHOS	1,2-bis[(2S,5S)-2,5-diisopropylphospholano]benzene

DTBM-SEGPHOS	[(4*R*)-(4,4′-bi-1,3-bezodioxole)-5,5′-diyl]bis[bis(3,5-di-tert-butyl-4-methoxyphenyl)phosphine]
DM-SEGPHOS	5,5’-bis[di(3,5-xylyl)phosphino]-4,4′-bi-1,3-benzodioxole
BIPHEP	bis(diphenylphosphino)-6,6′-dimethoxy-1,1′-biphenyl
XantPhos	4,5-bis(diphenylphosphino)9,9-dimethylxanthene
TADDOL	α,α,α,α-tetraaryl-1,3-dioxo-lane-4,5-dimethanol
BINOL	1,1′-binaphthalene-2,2′-diol
CSA	camphor sulfonic acid

Author details

Susana Porcel García
Institute of Chemistry, Universidad Nacional Autónoma de México,
Ciudad de México, México

*Address all correspondence to: sporcel@unam.mx

References

[1] Ito Y, Sawamura M, Hayashi T. Catalytic Asymmetric Aldol Reaction: Reaction of Aldehydes with Isocyanoacetate Catalyzed by a Chiral Ferrocenylphosphine-gold(I) complex. J. Am. Chem. Soc. 1986; 108:6405-6406. DOI: 10.1021/ja00280a056.

[2] Teles J H, Brode S, Chabanas M. Cationic Gold(I) Complexes: Highly Efficient Catalysts for the Addition of Alcohols to Alkynes. Angew. Chem. Int. Ed. 1998;37, 1415-1518. DOI: 10.1002/(SICI)1521-3773(19980605)37:10< 1415::AID-ANIE1415>3.0.CO;2-N.

[3] Hashmi A S K, Frost T M, Bats J W, J. Highly Selective Gold-Catalyzed Arene Synthesis. J. Am. Chem. Soc. 2000; 122, 11553-11554. DOI: 10.1021/ja005570d.

[4] Mizushima E, Kazuhiko S, Teruyuki H, Tanaka, M. Highly Efficient Au(I)-Catalyzed Hydration of Alkynes. Angew. Chem. Int. Ed. 2002; 41, 4563-4565. DOI: 10.1002/(SICI)1521-3773(20021202)41:23<4563::AID-ANIE4563>3.0.CO;2-U.

[5] Nevado C, Cárdenas D J, Echavarren A M, Reaction of Enol Ethers with Alkynes Catalyzed by Transition Metals: 5exo-dig vesus 6endo-dig Cyclizations via Cyclopropyl Platinum or Gold carbene Complexes. Chem. Eur. J. 2003; 9, 2627-2635. DOI: 10.1002/chem.200204646.

[6] Luzung M R, Markham J P, Toste F D. Catalytic Isomerization of 1,5-Enynes to Bicyclo[3.1.0]hexenes. J. Am. Chem. Soc. 2004;126, 10858-10859. DOI: 10.1021/ja046248w.

[7] Mamane V, Gress T, Krause H, Fürstner A. Platinum- and Gold-Catalyzed Cycloisomerization Reactions of Hydoxylated Enynes. J. Am. Chem. Soc. 2004;126, 8654-8655. DOI: 10.1021/ja048094q.

[8] Fürstner A, From Understanding to Prediction: Gold-and Platinum-Based Π-Acid Catalysis for Target Oriented Synthesis. Acc. Chem. Res. 2014;47:925-938. DOI: 10.102/ar4001789.

[9] Zhang Y, Luo T, Yang Z, Strategic Innovation in the Total Synthesis of Complex Natural Products Using Gold Catalysis. Nat. Prod. Rep. 2014;31, 489-503. DOI: 10.1039/C3NP70075E.

[10] Lo V K-Y, Chan A O-Y; Che C-M, Gold and Silver Catalysts: From Organic Transformation to Bioconjugation. Org. Biomol Chem. 2015; 13, 6667-6680. DOI: 10.1039/C5OB0407A.

[11] Dorel R, Echavarren A M, Gold(I)-Catalyzed Activation of Alkynes for the Construction of Molecular Complexity. Chem. Rev. 2015; 115, 9028-9072. DOI: 10.1021/cr500691k.

[12] Gao F, Zhou Y, Liu H, Recent Advances in the Synthesis of Heterocycles via Gold-Catalyzed Cascade Reactions: A Review. Curr. Org. Synth. 2017; 21, 1530-1566. DOI: 10.2174/1385272821666170102144500.

[13] Quach R, Furkert D P, Brimble M A, Gold Catalysis: Synthesis of Spiro, Bridged, and Fused Ketal Natural Products. Org. Biomol. Chem. 2017; 15, 3098-3104. DOI: 10.1039/C7OB00496F.

[14] Pirovano V, Gold-Catalyzed Functionalization Reactions of Indole. Eur. J. Org. Chem. 2018; 1925-1945. DOI: 10.1002/ejoc.201800125.

[15] Cheng X, Quintanilla C D, Zhang L, Total Synthesis and Structure Revision of Diplobifuranylone B. J. Org. Chem. 2019; 84, 11054-11060. DOI: 10.1021/aces.joc.9b01613.

[16] Gorin D J, Toste F D, Relativistic Effects in Homogeneous Gold Catalysis. Nature 2007; 446, 395-403. DOI: 10.1038/nature05592.

[17] Pyykkö P, Theoretical Chemistry of Gold. Angew. Chem. Int. Ed. 2004; 43, 4412-4456. DOI: 10.1002/anie.200300624.

[18] Rocchigiani L, Bochmann M, Recent Advances in Gold(III) Chemistry: Structure, Bonding, Reactivity, and Role in Homogeneous Catalysis. Chem. Rev. 2021; DOI: 10.2021/acs.Chemrev.0c00552. 10.3762/bjoc.9.264.

[19] López F, Mascareñas J L, Gold(I)-Catalyzed Enantioselective Cycloaddition Reactions, Beilstein J. Org. Chem. 2013; 9, 2250-2264.

[20] Zi W, Toste F D, Recent Advances in Enantioselective Gold Catalysis, Chem. Soc. Rev. 2016, 45, 4567-4589. DOI: 10.1039/c5cs00929d.

[21] Li Y, Li W, Zhang J, Gold-Catalyzed Enantioselective Annulations, Chem. Eur. J. 2017; 23, 467-512. DOI: 10.1002/chem.201602822.

[22] Jiang J J, Wong M K, Recent Advances in the Development of Chiral Gold Complexes for Catalytic Asymmetric Catalysis, 2021, DOI: 10.1002/asia.202001375.

[23] Muñoz M P, Adrio J, Carretero J C, Echavarren A M, Ligand Effects in Gold- and Platinum-Catalyzed Cyclizations of Enynes: Chiral Gold Complexes for Enantioselective Alkoxycyclization, Organometallics 2005; 24, 1293-1300. DOI: 10.1021/om0491645.

[24] Sethofer S G, Staben S T, Hung O Y, Toste F D, Au(I)-Catalyzed Ring Expanding Cycloisomerizations: Total Synthesis of Ventricosene, Org. Lett. 2008; 10, 4315-4318. DOI: 10.1021/ol801760w.

[25] Zheng H, Felix R J, Gagné M R, Gold-Catalyzed Enantioselective Ring-Expanding Cycloisomerization of Cyclopropylidene Bearing 1,5-Enynes. Org. Lett. 2014; 16, 2272-2275. DOI: 10.1021/ol500795l.

[26] Wu Z, Leboeuf D, Retailleau P, Gandon V, Marineti A, Voituriez A, Enantioselective Gold(I)-Catalyzed Rearrangement of Cyclopropyl-Substituted 1,6-Enynes into 2-Oxocyclobutyl-cyclopentanes. Chem. Commun. 2017; 53, 7026-7029. DOI: 10.1039/c7cc03234j.

[27] Kleinbeck F, Toste F D, Gold(I)-Catalyzed Enantioselective Ring Expansion of Allenylcyclopropanols, J. Am. Chem. Soc. 2009; 131, 9178-0179. DOI: 10.1021/ja904055z.

[28] García-Morales C, Ranieri B, Escofet I, López-Suarez L, Obradors C, Konovalov A I, Echavarren A M, Enantioselective Synthesis of Cyclobutenes by Intermolecular [2+2] Cycloadition with Non-C2 Symmetric Digold Catalysts, J. Am. Chem. Soc. 2017; 139, 13628-13631. DOI: 10.1021/jacs.7b07651.

[29] Cao Z-Y, Wang W, Liao K, Wang X, Zhou J, Ma J, Catalytic Enantioselective Synthesis of Cyclopropanes Featuring Vicinal All-Carbon Quaternary Stereocenters with a CH2F Group; Study of the Influence of C-F···H-N Interactions on Reactivity, Org. Chem. Front. 2018; 5, 2960-2968. DOI: 10.1039/c8qo00842f.

[30] Zheng Y, Guo L, Zo W, Enantioselective and Regioselective Hydroetherification of Alkynes by Gold-Catalyzed Desymmetrization of Prochiral Phenols with P-Stereogenci Centers, Org. Lett. 2018; 20, 7039-7043. DOI: 10.1021/acs.orglett.8b2982.

[31] Urbano A, Hernández-Torres G, Del Hoyo A M, Martínez-Carrión A, Carreño C, Mild Access to Planar-Chiral ortho-Condensed Aromatic Ferrocenes via Gold(I)-Catalyzed Cycloisomerization of ortho-Alkynylaryl Ferrocenes, Chem. Commun. 2016; 52, 6419-6422. DOI: 10.1039/c6cc0262a.

[32] Glinskt-Olivier N, Yang S, Retailleau P, Gandon V, Guinchard X, Enantioselective Gold-Catalyzed Picted-Spengler Reaction, Org. Lett. 2019; 21, 9446-9451. DOI: 10.1021/acs.orglett.9b03656.

[33] Zhang Z M, Chen P, Wenbo L, Niu Y, Zhao X L, Zhang J, A New Type of Chiral Sulfinamide Monophosphine Ligands: Stereodivergent Synthesis and Application in Enantioselective Gold(I)-Catalyzed Cycloaddition Reactions, Angew. Chem. 2014; 126, 4439-4443. DOI: 10.1002/ange.201401067.

[34] Zhou L, Xu B, Ji D, Zhang ZM, Zhang J, Ming-Phos/Gold(I)-Catalyzed Stereodivergent Synthesis of Highly Substituted Furo[3,4-*d*][1,2]oxazines, Chin. J. Chem. 2020; 38, 577-582. DOI: 10.1002/cjoc202000034

[35] Wang Y, Zhang Z-M, Liu F, He Y, Zhang J, Ming-Phos/Gold(I)-Catalyzed Diastereo- and Enantioselective Synthesis of Indolyl-Substituted Cyclopenta[c]furans, Org. Lett. 2018; 20, 6403-6406. DOI: 10.1021/acs.orglett.8b2701.

[36] Zhang P-C, Wang Y, Zhang Z-M, Zhang J, Gold(I)/Xiang-Phos-Catalyzed Asymmetric Intramolecular Cyclopropantion of Indenes and Trisubstituted Alkenes, Org. Lett. 2018; 20, 7049-7052. DOI: 10.1021/acs.orglett.8b02999.

[37] Cheng X, Wang Z, Quintanilla C D, Zhang L, Chiral Bifunctional Phosphine Ligand Enabling Gold-Catalyzed Asymmetric Isomerization of Alkyne to Allene and Asymmetric Synthesis of 2,5-Dihydrofuran, J. Am. Chem. Soc. 2019, 141, 3787-3789. DOI: 10.1021/jacs.8b12833.

[38] Zuccarello G, Mayans J G, Escofet I, Scharnagel D, Kirillova M S, Pérez-Jimeno A H, Calleja P, Boothe J R, Echavarren A M, Enantioselective Folding of Enynes by Gold(I) Catalysts with a Remote C_2-Chiral Element, J. Am. Chem. Soc. 2019; 141, 11858-11863. DOI: 10.1021/jacs.9b06326.

[39] Demmer C S, Voituriez A, Marinetti A, Catalytic Uses of Helicenes Displaying Phosphorous Functions, C. R. Chimie 2017; 20, 860-879. DOI: 10.106/j.crci.2017.04.002.

[40] Alonso I, Faustino H, López F, Mascareñas J L, Enantioselective Gold(I)-Catalyzed Intramolecular (4+3) Cycloadditions of Allendienes, Angew. Chem. 2011, 123, 11698-11702. DOI: 10.1002/ang.201105815.

[41] González A Z, Benitez D, Tkatchouk E, Goddard W A, Toste T D, Phosphoroamidite Gold(I)-Catalyzed Diastero- and Enantioselective Synthesis of 3,4-Substituted Pyrrolidines, J. Am. Chem. Soc. 2011; 133, 5500-5507. DOI: 10.1021/ja200084a.

[42] Teller H, Corbet M, Mantilli L, Gopakumar G, Goddard R, Thiel W, Fürstner A, One-Point Binding Ligands for Asymmetric Gold Catalysis: Phosphoramidites with TADDOL-Related but Acyclic Backbone, J. Am. Chem. Soc. 2012; 134, 15331-15342. DOI: 10.1021/ja303641pl.

[43] Liu B, Li K N, Luo S W, Huang J Z, Pang H, Gong L Z, J. Am. Chem. Soc. 2013; 135, 3323-3326. DOI: 10.1021/ja3110472l.

[44] Wang Y, Zhang P, Qian D, Zhang J, Highly Regio-, Diasterio, and Enantioselective Gold(I)-Catalyzed Intermolecular Annulations with N-Allenamides at the Proximal C=C Bond, Angew. Chem. 2015; 127, 15062-15065. DOI: 10.1002/ange.201507165.

[45] González-Fernández E, Nichoolls L D M, Schaaf L D, Farès C, Lehmann C W, Alcarazo M, Enantioselective Synthesis of [6]Carbohelicenes, J. Am. Chem. Soc. 2017; 139, 1428-1431. DOI: 10.1021/jacs.6b12443.

[46] Redero P, Hartung T, Zhang J, Nicholls L D M, Zichen G, Simon M, Golz C, Alcarazo M, Enantioselective

Synthesis of 1-Aryl Benzo[5]helicenes Using BINOL-Derived Cationic Phosphonites as Ancillary Ligands, Angew. Chem. Int. Ed. 2020; 59, 23527-23531. DOI: 10.1002/anie.202010021.

[47] Hand S, Slaugher L M, Enantioselective Alkynylbenzaldehyde Cyclizations Catalyzed by Chiral Gold(I) Acyclic Diaminocarbene Complexes Containing Weak Au-Arene Interactions, Angew. Chem. Int. Ed. 2012, 52, 2912-2915. DOI: 10.1002/anie.201107789.

[48] Zhang J Q, Lui Y, Wang X W, Zhang L, Synthesis of Chiral Bifunctional NHC Ligands and Survey of Their Utilities in Asymmetric Gold Catalysis, Organometallics 2019; 38, 3931-3938. DOI: 10.1021/acs.organomet.9b00400.

[49] Zhu X Z, Chamoreau L M, Zhang Y, Mouriès-Mansuy V, Fensterbank L, Bistri-Aslanoff O, Roland S, Sollogoub M, Permthylated NHC-Capped α- and β-Cyclodextrins (ICyDMe) Regioselective and Enantioselective Gold-Catalysis in Pure Water, Chem. Eur. J. 2020, 26, 15901-15909. DOI: 10.1002/chem.202001990.

[50] Hamilton G L, Kang E J, Mba M, Toste F D, sA Powerful Chiral Counterion Strategy for Asymmetric Transition Metal Catalysis, Science, 2007; 317, 496-499. DOI: 10.1126/science.1145229.

[51] Zi W, Toste F D, Gold(I)-Catalyzed Enantioselective Desymmetrization of 1,3-Diols through Intramolecular Hydroalkoxylation of Allenes, Angew. Chem. Int. Ed. 2015; 54, 14447-14451. DOI: 10.1002/anie.201508331.

[52] Tu X F, Gong L Z, Highly Enantioselective Transfer Hydrogenation of Quinolines Catalyzed by Gold Phosphates: Achiral Ligand Tuning and Chiral-Anion Control of Stereoselectivity, Angew. Chem. Int. Ed. 2012, 51, 11346-11349. DOI: 10.1002/anie.201204179.

[53] Shinde V S, Mane M V, Vanka K, Mallick A, Patil N T, Gold(I)/Chiral Bronsted Acid Catalyzed Enantioselective Hydroamination-Hydroarylation of Alkynes: The Effect of a Remote Hydroxyl Group on the Reactivity and Enantioselectivity, Chem. Eur. J. 2015; 21, 975-979. DOI: 10.1002/chem.201405061.

[54] Cala L, Mendoza A, Fañanás F J, Rodríguez F, A Catalytic Milticomponent Coupling Reaction for the Enantioselective Synthesis of Spiroacetals, Chem. Commun. 2013; 49, 2715-2717. DOI: 10.1039/c3cc00118k.

[55] Zhang Z, Smal V, Retailleau P, Voituriez A, Frison G, Marinetti A, Guinchard X, Tehtered Counterion-Directed Catalysis: Merging the Chiral Ion-Pairing and Bifuncitonal Ligand Strategies in Enantioselective Gold(I) Catalysis, J. Am. Chem. Soc. 2020; 142, 3797-3805. DOI: 10.1021/jacs.9b11154.

[56] Bohan PT, Toste F D, Well-Defined Chiral Gold(III) Complex Catalyzed Direct Enantioconvergent Kinetic Resolution of 1,5-Enynes, J. Am. Chem. Soc. 2017, 139, 11016-11019. DOI: 10.1021/jacs.7b06025.

[57] Reid J P, Hu M, Ito S, Huang B, Hong C M, Xiang H, Sigman M S, Toste F D, Strategies for Remote Enantiocontrol in Chiral Gold(III) complexes Applied to Catalytic Enantioselective γ,δ-Diels-Alder Reactions, Chem. Sci. 2020; 11, 6450-6456. DOI: 10.1039/d0sc00497a.

[58] Jian J J, Cui J F, Yang B, Ning Y, Lai N C H L, Wong M K, Chiral Cyclometalated Oxazoline Gold(III) Complex-Catalyzed Asymmetric Carboalkoxylation of Alkynes, Org. Lett. 2019, 6289-6294. DOI: 10.1021/acs.orglett.9b02171.

Chapter 4

Chiral Alkaloid Analysis

Ngoc Van Thi Nguyen, Kim Ngan Huynh Nguyen, Kien Trung Nguyen, Kyeong Ho Kim and Hassan Y. Aboul-Enein

Abstract

Alkaloids are distributed in plant kingdom and play important role in protection, germination as well as plant growth stimulants. Most of them are chiral compounds and are clinically administered as the racemic mixture, even though its enantiomers have been known to exert different pharmacological activity. Liquid chromatography using chiral stationary phases (CSP) proved to be an essential tool with a wide range of applications, including analysis of the stereochemistry of natural compounds. This review gives an overview of chiral separation alkaloids that were used in theoretical studies and/or applications in recent years. It shows the possibilities of polysaccharide CSPs have now also been established as the first-choice of chiral phases for enantiomer separation.

Keywords: chiral alkaloid, sample preparation, enantiomer separation, chiral stationary phase

1. Introduction

Over the centuries, humans have depended on nature for their essential needs of food supplies, shelters, apparel, transport means, fertilizers, flavors and fragrances, and the last not but least, medicines. Sophisticated traditional medicine systems have been generated by the plants over thousands of years. Moreover, plants maintain the significant sources of modern remedies for humanity. Additionally, according to WHO, 80% of the world's population—primarily those of developing countries rely on plant-derived medicines for their healthcare [1]. People continue to consider nature as a source of potential chemotherapeutic agents. Over 50% of clinical drugs all over the world are the product of natural plants and their derivatives. While more than 25% of the total are extracted from higher plants [2].

History of pharmacy was for centuries identical with the history of pharmacognosy, or study of materia medica, which were obtained from natural sources—mostly plants but minerals, animals, and fungi. Chirality is one of the universal phenomena in nature. For instance, chiral biomolecules such as amino acids, sugars, proteins and nucleic acids have created living organisms. In natural surroundings, these biomolecules are present in one of the two possible enantiomeric forms, e.g., amino acids in the L-form and sugars in the D-form. Living organisms show variation in biological responses to one of a couple of enantiomers in medicines due to the chirality [3].

A range of chemicals that accurate enzymatic metamorphosis defines stereochemical configurations. Consequently, there is a certain chirality in most organic compounds in nature. It is important to emphasize that some phytochemicals exist in only one enantiomeric form, while others the optical rotation of the metabolite can be different [4].

2. Background on chiral alkaloid

Alkaloids are cyclic organic compounds that contain nitrogen in a negative oxidation state. They are generally distributed in flora and are an essential role in plant protection, sprouting and stimulating plant growth. Alkaloids-containing plants are often used as traditional medicines and these compounds usually have marked pharmacological activity [5]. Over 21,000 alkaloids have been identified, which thus constitute the largest group among the nitrogen-containing secondary metabolites [6]. Alkaloids are significantly pharmaceutical, e.g. morphine as pain relief medicines, codeine as an antitussive in cough medicines, colchicine in the treatment of gout and familial Mediterranean fever (FMF), Quinine as an anti-malarial and a muscle relaxant, Quinidine, as an antiarrhythmic agent to prevent ventricular arrhythmias and L-hyoscyamine (in the form of its racemic mixture known as atropine) as antimuscarinic; i.e., as an antagonist of muscarinic acetylcholine receptors [7].

The first isolations of alkaloids in the nineteenth-century new investigation into medicine of several alkaloid-containing drugs and were accidental with the advent of the separation process for the extraction of drugs. In 1803, the French apothecary Derosne probably isolated narcotine. Several years later, the Hanoverian apothecary Sertürner further investigated opium (1806) and isolated morphine (1816) [7].

Based on their structures, alkaloids are divided into several subgroups: non-heterocyclic alkaloids and heterocyclic alkaloids, which are again divided into 7 major groups according to their basic ring structure [8]. Families reported to be rich in alkaloids are: Liliaceae, Amaryllidaceae, Apocynaceae, Berberidaceae, Leguminosae, Papaveraceae, Ranunculaceae, Rubiaceae and Solanaceae [9]. Most of alkaloids are chiral compounds and are clinically administered as the racemic mixture, although its enantiomers have been shown to exert different pharmacological activity.

2.1 Non-heterocyclic alkaloids

Phenylethylamine alkaloids in medicinal herbs (i.e. Citrus species and *Ephedra sinica*) are used ubiquitously for their effects on the metabolic process of humans by stimulating lipolysis and thus supporting to reduce the fat mass in obese people. Particularly, Ephedra Herba (Ma Huang) contain several alkaloids such as (1R, 2S)-(−)-ephedrine, (1S, 2S)-(+)-pseudoephedrine, (1R, 2S)-(−)-norephedrine, (1S, 2S)-(+)-norpseudoephedrine, (1R,2S)-(−)-N-methylephedrine, and (1S, 2S)-(+)-N-methylpseudoephedrine [10]. Each of these six compounds also has an enantiomer that does not occur naturally in the plant [11, 12]. Separation and quantification of optical isomers of ephedrine-type alkaloids are important since ephedrine-type alkaloids in natural have been found to be strengthened with inexpensive (racemic) synthetic similarity, and these enantiomers could exhibit important differences in pharmacological activities. To diminish essential public health risk, adulteration of Ephedra products could be discovered by the presence of both enantiomers, such as naturally occurring (−)-ephedrine and synthetic (+)-ephedrine in the samples [13].

In the case of C. urantium alkaloids, synephrine has also effect on human metabolism that could help to reduce fat mass in obese people, since it stimulates lipolysis, raises the metabolic rate and promotes the oxidation of fat through increased thermogenesis [14]. Synephrine is a chiral compound and is clinically administered as the racemic mixture, although its enantiomers have been illustrated to apply different pharmacological activity on α- and β-adrenoreceptors. Particularly, (R)-(−)-synephrine is from 1 to 2 orders of magnitude more active than its (S)-(+)-counterpart (**Figure 1**) [15].

2.2 Heterocyclic alkaloids

2.2.1 Tropane alkaloids

Solanaceae contain mainly tropane alkaloids such as atropine, anisodamine and scopolamine; these plants are extensively used both in traditional medicine and as sources for the extraction of the pharmacologically important (parasympatolytic and anti-cholinergic) alkaloids [10]. Atropine is existed in racemic mixture of (S)-hyoscyamine and (R)-hyoscyamine. (S)-hyoscyamine is original in plants and (R)-hyoscyamine forms under alkaline conditions. (S)-hyoscyamine functions competitive antagonist of muscarinic receptors, thereby inhibiting the parasympathetic activities of acetylcholine on the salivary and sweat glands, as well as gastrointestinal tract, while the (R)-hyoscyamine is mostly inactive. Atropine, which is more often applied than (S)-hyoscyamine, exhibits approximately half of the pharmacological activity of (S)-hyoscyamine. In reverse, Scopolamine is mostly applied as pure enantiomer, e.g. (S)-scopolamine bromide [16].

Anisodamine, a tropane alkaloid isolated from Solanaceae family (Scopolia tangutica Maxim.). In China for decades, Anisodamine is an effective cholinoceptor antagonist and has been used as a spasmolytic drug to effect on smooth muscle by feature of its weaker side effect on the central nervous system than atropine. This kind of alkaloids have biological characteristic including cholinoceptor agonists and antagonists, like most chiral drugs, depend strongly on their stereochemistry. The effectiveness differences among four isomers of anisodamine racemic on muscarinic receptors have been perceived (**Figure 2**) [17].

2.2.2 Aconitine alkaloids

Aconitum plants (Ranuncolaceae) are generally distributed across Asia and North America. In the Chinese Pharmacopeia, two species of them, A. carmichaeli Dexb. and A. kusnezoffiiare were listed. Aconitine and the congener mesaconitine and hypaconitine (**Figure 3**) are the important diester-diterpene alkaloids of aconitum plants. Although they have toxic effects on human health, they can also be used at low doses because their pharmacological effects such as anti-inflammatory and anti-pain are effectively [10].

2.2.3 Quinolizidine alkaloids

In the legume alkaloids, the largest single group is quinolizidine alkaloids. In distribution to species in the more primitive tribes of the Papilionoideae, they appear to be restricted. Because of their toxicity in humans and animals as

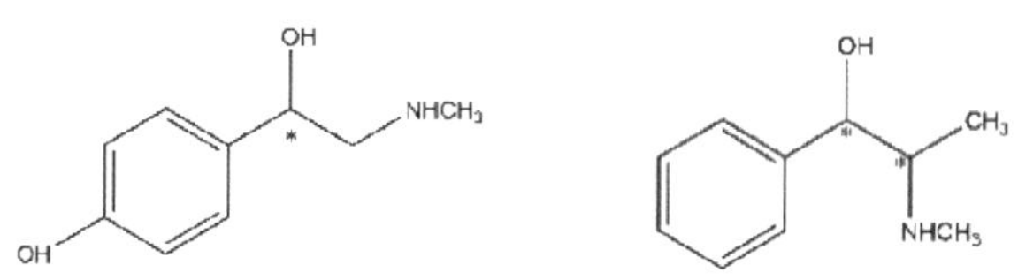

Figure 1.
*Molecular structure of synephrine and ephedrine alkaloids (*chiral center).*

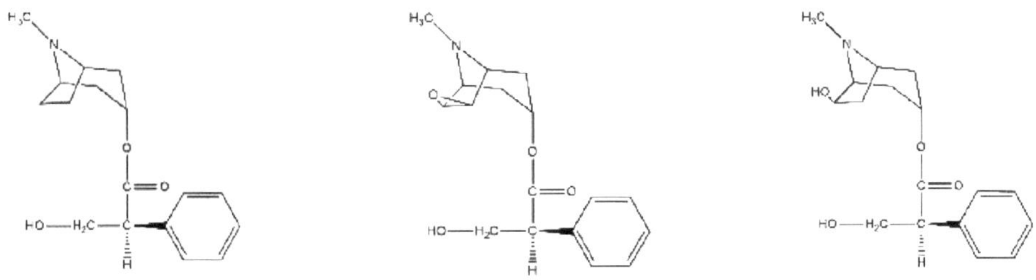

Figure 2.
Molecular structure of (S)-hyoscyamine, (S)-scopolamine, and anisodamine.

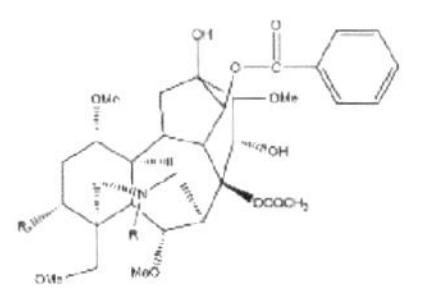

Alkaloid	R1	R2
Aconitine	C2H5	OH
Mesaconitine	CH3	OH
Hypaconitine	CH3	H

Figure 3.
Molecular structure of aconitine, mesaconitine and hypaconitine.

components of poisonous plants, these compounds become important. In contrast, some of them are potentially useful in pharmacological activities [18]. Radix sophorae flavescentis (*Sophora flavescens*) is frequently used in Traditional Chinese medicine for treating acute hepatitis and jaundice; it was found that quinolizidine alkaloids were the main constituents of this herbal drug such as matrine, sophoridine, sophocarpine, lehmannine, sophoranol, oxy-matrine, oxysophocarpine which have some chiral center [10, 19]. In natural, they exist as an isomer, sophocarpine (**Figure 4**) is an example. The naturally (−)-sophocarpine isolated from the root of the Chinese medicinal herb *Sophora flavescens* Ait. (Fam. Leguminosae) [20].

2.2.4 Isoquinoline alkaloids

Bis-benzylisoquinoline alkaloids have fascinated by the significant pharmacological impacts; especially, protoberberines are a structural class of organic cations (quaternary ammonium alkaloids) mostly distributed in Ranuncolaceae (e.g., Rhizoma coptidis), Berberidaceae (e.g. Cortex berberdis), Papaveraceae (e.g. Herba chelidoni) and Rutaceae (e.g. Cortex phellodendri) [10]. The most considered chiral isoquinoline alkaloids are tetrahydroprotoberberine backbone structure such as tetrahydropalmatine (THP), tetrahydroberberine (THB), and corydaline [21]. (DL)-THP and (DL)-THB are highly abundant in C. yanhusuo and a variety of Corydalis plants. (L)-THP can also be isolated from Stephania plants [22]. Tetrahydropalmatine is one of the active ingredients isolated from Rhizoma Corydalis (yanhusuo), a traditional Chinese medicine that has been used for the treatment of chest pain, epigastric pain, dysmenorrhea, traumatic swelling, and pain for thousands of years [23]. The analgesic activity of (−)-THP is much higher than that of (+)-THP. Clinically, THP is used as the racemic mixture (**Figure 5**) [22].

Amaryllidaceae alkaloids are an important class of iso-quinoline derivatives; among them galanthamine, that is found in Galanthus and Narcissus species, has been approved for the pharmacological treatment of Alzheimer's disease [24]. There are several chiral centers in this molecule, but only one S-enantiomer responsible for Alzheimer's disease, other stereoisomers considered as impurity (**Figure 6**) [25].

Morphinane alkaloids (opium alkaloids) such as morphine, codeine, thebaine, papaverine and narcotine belong to isoquinoline derivatives and show a broad

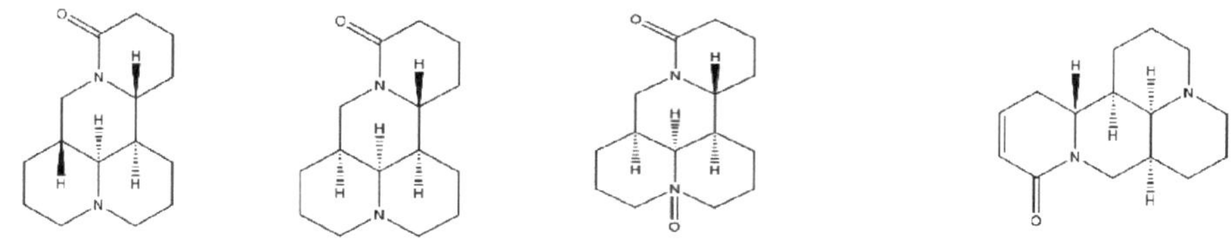

Sophoridine Matrine Oxymatrine (-) Sophocarpine

Figure 4.
Molecular structures of four quinolizidine alkaloids.

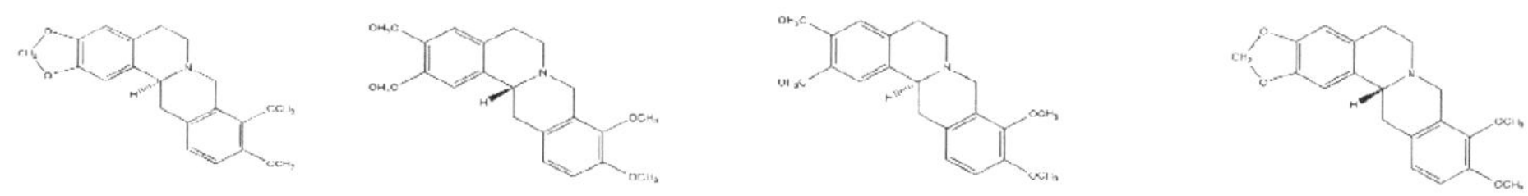

(-)-THP (+)-THP (-)-THB (+)-THB

Figure 5.
Molecular structures of the enantiomers of tetrahydropalmatine (THP), and tetrahydro-berberine (THB).

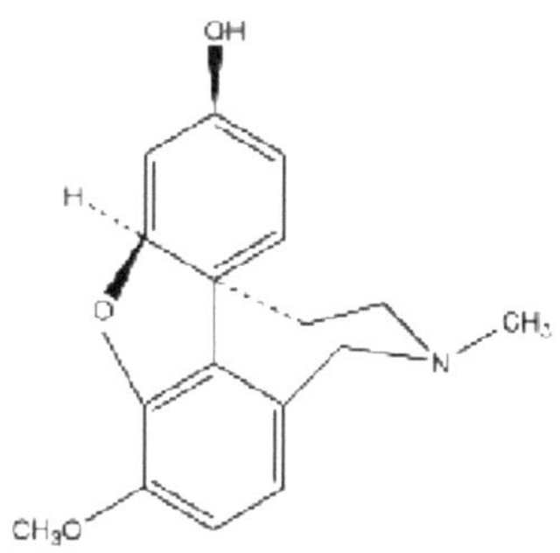

Figure 6.
Molecular structures of galanthamine.

range of pharmacological activities; their major application is in analgesia, sedation and cough depression [26]. Although opiate alkaloids have an important place in medicine, the illegal trafficking and abuse of heroin (the diacetyl derivative of morphine) has become a widespread problem [27]. Opium, the exudates from *Papaver somniferum*, contains more than 30 alkaloids and is the raw material for extraction; also, the dried heads of *P. somniferum*, so-called poppy straw, is used as a source of morphine and thebaine. **Figure 7** show the molecular structure of opium alkaloids, most of them have multi chiral center.

2.2.5 Pyrrolizidine alkaloids

Although these alkaloids have at present no great medicinal significance, they are important in that they constitute the poisonous hepatotoxic constituents of plants of the genus Senecio (Compositae), well-known for their toxicity to livestock [28]. Pyrrolizidine alkaloids are found in a variety of plant species growing wide world such as Gynura segetum that belongs to the Compositae family and Senecio and Tussilago genera [10]. The majority of naturally occurring pyrrolizidine alkaloids (PA) are hepatotoxic causing liver damage and in some cases liver cancer. Toxic PAs are often responsible for serious health problems through direct consumption of PA-containing herbal teas, herbal medicines, and herbal dietary supplements [29].

The most important pyrrolizidine alkaloids senecionine, seneciphylline, retrorsine and senkirkine, contain the 4-azabicyclo [3.3.0] octane system with senecionine and seneciphylline differing only for the presence in the latter of the C_{13}-C_{23} double bond (**Figure 8**) [30].

2.2.6 Indole alkaloids

Indole alkaloids constitute a wide class of natural products most of them pharmacologically important and characterized by very different activities [31]. In the recent years, attention has been focused on the biological activity of yohimbine which is a monoterpenoid indole alkaloid (**Figure 9**). It displayed the treatment of erectile functional disturbance and anxiogenic [32]. Hydroindole alkaloids such as mesembrine and congeners (mesembrenone, Δ_7 mesembrenone, mesembranol and its stereoisomer epimesembranol) have been isolated from Sceletium species used for the psychoactive effects [33].

The vinca alkaloids were isolated from the Madagascar periwinkle, Catharantus roseus G. Don., which included a class of about 130 terpenoid indole alkaloids [32]. In early 1965, people obviously know their clinical quality. And this group of compounds has been taken advantage of as an anticancer servant for more than 40 years and is a symbol of the compound that gives the trend to drug development [34, 35]. Among these base (+)-vincamine exhibits a valuable therapeutic activity in cerebral insufficiencies. Due to the presence of three stereogenic centers eight stereoisomers (four enantiomeric pairs) are in fact possible (**Figure 10**) [36].

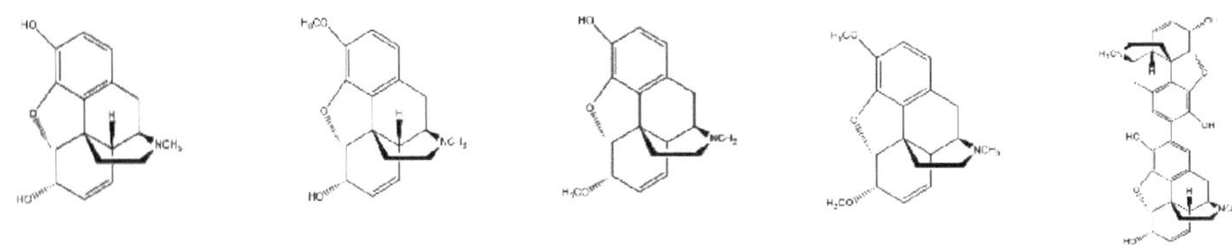

(1) (2) (3) (4) (5)

Figure 7.
Molecular structures of: (1) morphine; (2) codeine; (3) oripavine; (4) thebaine; (5) pseudomorphine.

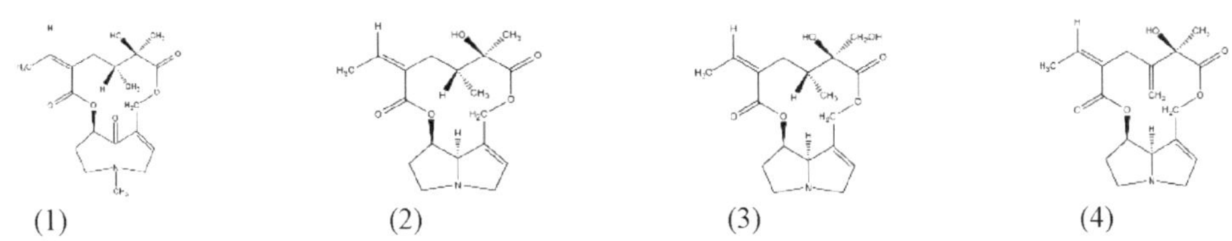

(1) (2) (3) (4)

Figure 8.
Molecular structures of four toxic PAs. (1) senkirkine; (2) senecionine; (3) retrorsine; (4) seneciphylline.

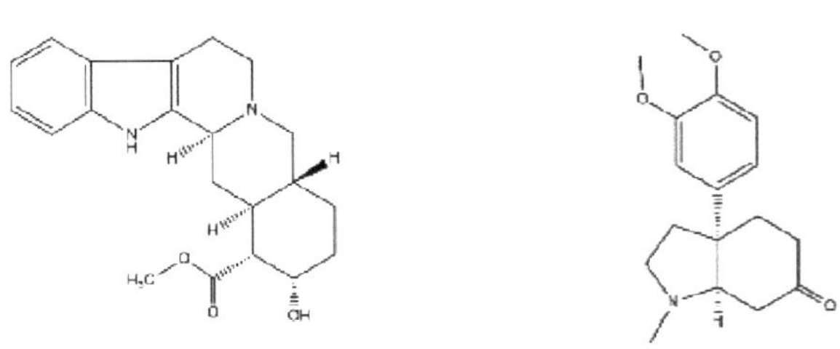

Figure 9.
Molecular structure of yohimbine and mesembrine.

2.2.7 Miscellaneous alkaloids

Steroidal alkaloids including verticine and verticinone are distinguished by cholestane carbon skeleton (isosteroid alkaloids) with a hexacyclic benzo [7, 8] fluoreno [2,1-b] quinolizine nucleus (**Figure 11**). These compounds have been isolated from plants from Liliaceae family typically Bulbus fritillariae used as a traditional medicine in Japanese, Turkish, Pakistani, and south-east Asian folk medicines [10]. Pharmacological studies demonstrate that verticine and verticinone in Bulbus Fritillariaeare the primary active ingredients responsible for the antitussive activity [37].

Stemona, belonging to Stemonaceae family, is known in the folk medicine of Southeast Asia, China, and Japan since its Phyto-preparations (primary the roots) are used to treat diseases about bronchitis, pertussis and tuberculosis. Interestingly many alkaloids, structurally defined as pyrido[1,2-α]azepines, have been recognized in this plant species and are considered the important pharmacological activity. All the Stemona alkaloids are polycyclic and contain multiple stereocenters [38]. Up to now, there are about 139 Stemona alkaloids which the scientist isolated (**Figure 12**).

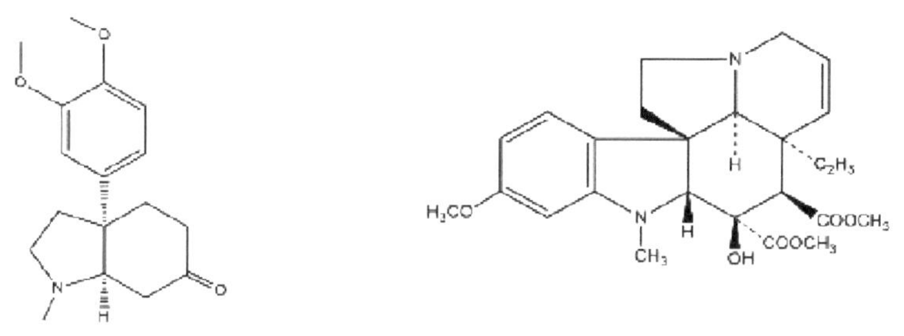

(a) (b)

Figure 10.
Molecular structure of major vinca alkaloids isolated from Catharanthus roseus*: (+)-Catharanthine (A) and (−)-Vindoline (B).*

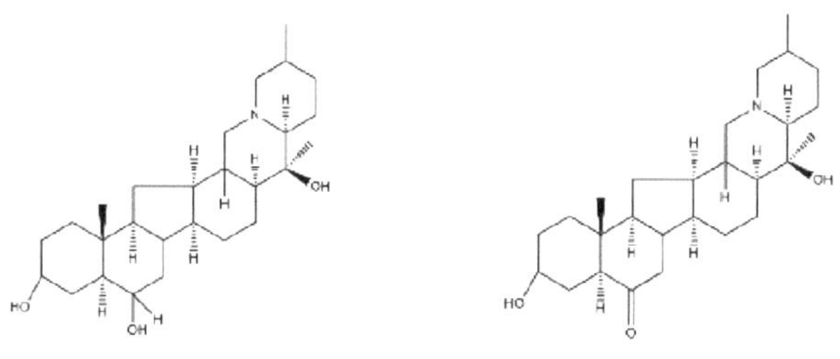

Figure 11.
Molecular structures of verticine and verticinone.

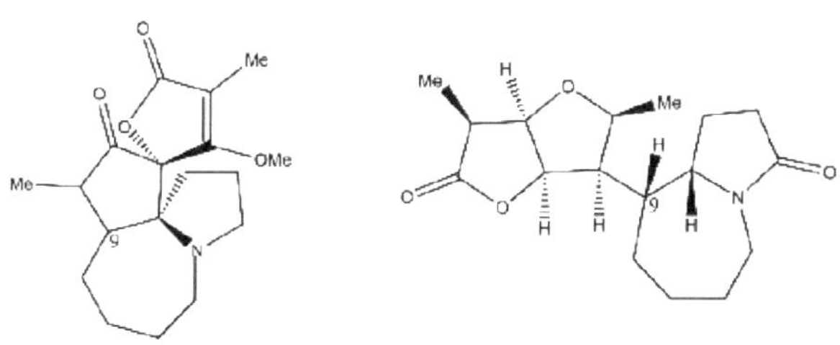

Figure 12.
Molecular structure of stemonamine and parvistemoline.

3. Techniques of extraction and purification

Analytical methods usually contain several steps, such as sampling, sample preparation, isolation, and quantification. Remarkably sample preparation just recently is concentration as an important analytical step. According to Wen *et al*. [39, 40], the main objective of sample preparation are removal of interferences, and preconcentration of the analytes that is considered a bottleneck of analytical processes. In this chapter, we will provide the current state of the art in sample preparation for analyzing alkaloids in herbal matrices, focusing on extraction, clean-up steps and purification.

3.1 Extraction

3.1.1 Ultrasound assisted extraction (UAE)

Ultrasound Assisted Extraction (UAE) technique is based on the using of acoustic waves in the kilohertz range spreading in liquid medium. These waves created by the ultrasound produces regions of compression and rarefaction in the molecules. Then, the cavitation bubbles are formed and collapse giving rise to smaller bubbles that could act as new cavitation nuclei or simply get dissolved. When the bubbles collapse at the surface of the herbal material, a shockwave having very high temperature and pressure is induced, resulting in plant cell disruption which enhances both the mass transfer of alkaloids into solution and the solvent penetration [41]. In addition, the swelling of plant materials can be enhanced by ultrasound, leading to improving of solvent penetration which increases the extraction yield [42]. The UAE procedure is optimized with regard to extraction solvent, temperature and liquid to solid ratio for the plant sample [41]. UAE has some advantageous properties including high extraction efficiency, good reproducibility, low solvent consumption, low cost and environmental friendliness. However, the major disadvantage of UAE is generating heat, leading to the degradation of thermo labile products and racemization of chiral compounds [43]. To avoid such types of drawback, extraction is carried on under an ice bath to reduce the temperature [44]. **Table 1** lists examples of protocols that were developed using MAE from various plants.

3.1.2 Microwave assisted extraction (MAE)

The MAE technique uses the electromagnetic radiations with a frequency range of 0.3–300 GHz, that stimulates ion migration and dipole rotation leading to the heating of dielectric materials and the penetration of extraction solvent into the matrix [49]. The released thermal energy is increased gradually with higher dielectric constant, so the effectiveness of MAE is depended on the dielectric properties of both extraction solvent and sample matrix [50]. Therefore, only specific solvents which have high dipole moment as water, methanol and ethanol can be used for extraction solvent in MAE and the moisture of plant sample is an important factor in the extraction efficiency [51]. Basically, higher water content matrices will be expected to give higher extraction yields. Water contained in plant matrices absorbs microwave radiation creating pockets of localized heating in the sample. This heat promotes the plant cell walls rupture which enhances the release of alkaloids into the solvent and the increase of extraction yield [50]. In addition to having a high dielectric constant, the extraction solvents must have a high dissipation factor to reduce the potential of localized sample overheating resulting degradation of alkaloids in plants [43]. Therefore, organic solvent-water mixtures, polar

Plant	Compounds	Solvent	Solvent: biomass (mL: g)	Temp (°C)	Time duration (min)	Ref.
Macleaya cordata	Protopine Allocryptopine Sanguinarine Chelerythrine Dihydrochelerythrine Dihydrosanguinarine	1-hexyl-3-methylimidazolium tetrafluoroborate ([C6MIM][BF4]) aqueous solution	500: 1	80	15	[45]
Catha edulis	Norephedrine Cathine Cathinone	0.1 N HCl	600: 1	—	45	[46]
Ipomoea genera	Ergot alkaloid (ergine, ergometrine, lysergic acid α-hydroxyethylamide) Penniclavine Chanoclavine Lysergol	70% methanol	100: 1	60	30	[47]
Carica papaya	Carpaine Pseudocarpaine Dehydrocarpaine I Dehydrocarpaine II	100% methanol	100: 7.5	—	20	[48]

Table 1.
Extraction conditions from various plants using UAE.

organic solvents and water are usually used as extraction solvent. Moreover, other parameters relating to extraction performance as sample size, sonication power, solid to liquid ratio, extraction time and microwave power should be modified for MAE procedure optimization [34]. The main advantages of MAE are the low solvent consumption, the ability to extract many samples simultaneously and the short extraction time [43, 52]. The major drawbacks of MAE are nonhomogeneous heating distribution and overheating of extract which may cause racemization or thermal degradation of chiral alkaloids [44]. **Table 2** lists examples of protocols that were developed using MAE from various plants.

3.1.3 Supercritical fluid extraction (SFE)

The SFE process utilizes pressurized fluids (mainly CO_2) as extraction solvents. In this technique, a fluid is heated and compressed to reach above critical point of it's creating the fluid having physicochemical properties of both liquid and gas states called supercritical fluid [58]. Specifically, supercritical fluid has a density similar to liquid (0.3–0.8 g/cm^3), a viscosity similar to gas (10^{-4} – 10^{-3} g/s.cm) and a diffusion coefficient that is intermediate between liquid and gas [59]. Therefore, supercritical fluid has higher transport capacity which facilitate to fluid diffusion through plant materials in comparison to traditional extraction solvents [43]. In addition, the density is related to polarity property of fluid which directly impact in solubility of compounds in extraction solvent. This parameter can be modified by controlling temperature and/or pressure so the flexibility and selectivity of the technique is enhanced, enabling selective extraction of different compounds from the plant matrix [60]. Carbon dioxide is commonly used in SFE because it has ideal properties including low critical temperature (31.3°C) and can be easily remove from extracts [51]. However, carbon dioxide is less effective in extraction of polar compounds from matrix because of its low polarity property. Aiming to extract

Plant	Compounds	Solvent	Solvent: biomass (mL: g)	Tem (°C)	Time duration	Micro wave power (W)	Ref.
Peganum harmala	Vasicine, Harmalin Harmine	80% ethanol	30: 1	80	8 min	600	[53]
Stephania sinica	Sinoacutine Palmatine Isocorydine L-tetrahydro palmatine	65% ethanol	24: 1	60	90 s	150	[54]
Lotus plumule	Liensinine Dauricine Isoliensinine Neferine Nuciferine	65% methanol	—	—	4 min	200	[55]
Corydalis decumbens	Protopine Palmatine Allocryptopine Jatrorrhizine Tetrahydro palmatine Corypalmine Bicuculline	90% methanol	20: 1	40	5 min	—	[56]
Menispermum dauricum	Bianfugedine, Menisporphine, 6-O-demethyl menisporphine Bianfugecine Dauriporphine Dauriporphinoline	70% ethanol	20: 1	60	11 min	—	[57]

Table 2.
Extraction conditions from various plants using MAE.

more polar alkaloids, the modifiers such as methanol, ethanol or water are added to extend the range of the solvating strength [43, 47]. For optimization of SFE procedure, these parameters such as pressure, temperature, modifier, flow of carbon dioxide and modifier [51]. Besides, the extraction time also effects on extraction yield, since an inadequate extraction time can result in incomplete extraction, while too long extraction time can cause the degradation of compounds. The major drawback of SFE are the complexity of system configuration and the requirement for a personal training program to operate the instrument [61].

3.1.4 Pressurized solvent extraction (PSE)

PSE process uses the pressurize solvents to enhance transport capacity of solvents and mass transfer rates which leads to improve extraction performance. In this technique, extraction solvents are heated at/or above the solvent's boiling points to decrease viscosity while keeping its in liquid state thanks to an elevated pressure [62]. Therefore, the extraction process is enhanced kinetic which leads to decreasing both the extraction time and solvent consumption. Similar to SFE, these parameters such as solvents nature, temperature and pressure should be modified to optimize PSE procedure [51]. Logically, higher temperature would be expected to give higher alkaloid extraction yields. However, excessive temperature may cause degradation and racemization of chiral alkaloids. The main advantages of PSE are utilizing an extensive range of solvents (except strong acids/bases), low solvent consumption, short extraction time, automated instruments and performing an oxygen- and light – free extraction condition. Besides, the major drawbacks of PSE are similar to SFE such as using expensive laboratory equipment and requirement for a professional training to operate instrument [43].

3.2 Purification

Due to the alkaloids usually exist in plants at low concentration and the complication of plant matrices, samples should be purified and enriched which facilitate to identify and/or quantify process right after extraction step. Liquid–liquid extraction (LLE) and solid phase extraction (SLE) are popular clean-up methods utilized for sample preparation of alkaloids. Moreover, other techniques based on LLE and SPE methods, such as Liquid Membrane Extraction (LME) and Solid-Phase Micro Extraction (SPME) have also been developed.

3.2.1 Liquid–liquid extraction (LLE)

Liquid–Liquid Extraction (LLE) is the most simple and traditional clean-up method. LLE method is based on the relative solubility of compounds between two immiscible solvents. The alkaloids have polarity varying between pH, so the solubility of alkaloids in specific solvent are also affected by pH [34]. In the acid solutions (pH is lower than pKa of alkaloids), the analytes are protonated which leads to better water solubility so this aqueous phase can be washed with less polar organic solvents such as ethyl acetate, n-hexane and diethyl ether to eliminate hydrophobic interferences. After that, this aqueous layer is alkalinized which leads to the alkaloids becoming non-polarity and can be extracted by organic solvents to eliminate hydrophilic interferences [63]. For enrichment, the organic solvents layer can be collected, vaporized and reconstituted into new solvents which is suitable for analytical instrument. The main disadvantages of this method are requirement for repetitive extraction causing time consuming and solvent wasting [51].

3.2.2 Solid phase extraction (SPE)

To overcome the drawback of LLE method, solid phase extraction (SPE) has been developed and applied in sample preparation since the 1970s. In this method, extract is loaded onto a sorbent phase which will retain alkaloids. Then, interferences in extract are washed away and the analytes is eluted by suitable solvents [64]. In fact, cartridge is the most popular SPE type due to its convenience. The SPE procedure have five step including conditioning, loading, washing and elution step. Several factors can affect to the extraction efficiency such as concentration of analytes in solvent, loading solvent nature, sorbent types, particle size, volume used for loading – washing – eluting, flow rate and elution solvent. Each factor has specified role which depends on the affinity of analytes and solid phase. Because the chemical structures of alkaloids always have secondary or tertiary amine groups, the strong cation exchange (SCX) sorbents are an ideal choice for sample preparation. When using these sorbents, washing solvent will be water and organic solvent to eliminate both hydrophilic and hydrophobic from plant matrices. After that, alkaloids will be deprotonated for elution by alkalized solvent which has pH at 2 units above pKa of analytes and evaporated to enrich sample [64].

If the analytes are unstable in strong alkaline solutions, the weak cation exchange (WCX) will be use instead. The WCX sorbent has carboxylic acid as functional group which has pKa value about 4.8, so these sorbents should be conditioned by solutions having pH above 6.8 for sorbent ionization. In addition, the loading and washing solvent pH should be adjusted at the value above 6.8 and below 2 value of analytes's pKa to maintain the ionized state of both sorbent and analytes. Finally, the alkaloids will be eluted by the acidic solutions [63]. Besides, the C_{18} and C_8 sorbents are also applied to extract aromatic alkaloids and eliminate hydrophilic interferences from matrices. In some case, those sorbents could be used for eliminating hydrophobic interferences by loading unretained alkaloids through cartridges [64].

4. High performance liquid chromatography in chiral separation

High performance liquid chromatography (HPLC) is currently the most widely used chromatographic enantio-separation technique [65]. HPLC has become one of the most common modern chemical analysis techniques because of its versatility, efficiency, stability, reproducibility and sensitivity. With these advantages, HPLC continues to be one of the best choices for chiral analysis and separation. Basically, chiral separation by HPLC techniques included direct and indirect methods. The indirect method is based on diastereomer formation by the derivatization reaction of analytes and a chiral reagent, then the separation of diastereomeric derivatives is performed by using a column having an achiral stationary phase. In addition, the direct method uses a chiral stationary phase for chiral separation or forms diastereomer by using a chiral mobile phase additive (CMPA).

HPLC using CSPs has demonstrated to be extremely useful, accurate, versatile, and it has been a widely used technique in diverse fields and applications, emphasizing (**Table 3**). The CSP mode is generally the most straightforward and convenient means for chromatographic enantiomer separation; it is the method of choice for both analytical and preparative applications [66–69]. A hundred CSPs have been developed and commercialized thirty years ago [70]. Besides, the larger number of CSPs are made in laboratory for specialized separation. CSPs are divided into nine major types by Snyder basing on the interaction mechanism between stationary phase and analytes [71].

- Macromolecular selectors of semisynthetic origin (polysaccharides)

- Macromolecular selectors of synthetic origin (poly(meth)acrylamides), (poly-tartramides)

- Macromolecular selectors of natural origin (proteins)

- Macrocyclic oligomeric or intermediate-sized selectors (cyclodextrins, macrocyclic antibiotics, chiral crown ethers)

- Synthetic, neutral entities of low molecular weight (Pirkle-type phases, brush-type CSPs)

- Synthetic, ionic entities of low molecular weight that provide for ion exchange

- Chelating selectors for chiral ligand-exchange chromatography.

Type	CSP	Typical column trade name	Application
I	Polysaccharide	AD, OD, OJ, AS, IA, IB, IC	Alkaloids, tropines, amines, beta blockers, aryl methyl esters, aryl methoxy esters
II	Synthetic-Polymer CSPs	Kromasil CHI-DMB and CHI-TBB	Acidic, neutral, and basic compounds
III	Protein Phases	Chiral HSA, Chiral AGP, Ultron ES-OVM, Chiral CBH	Benzodiazepine, Warfarin and oxazepam, beta blockers
IV	Cyclodextrin	Cyclobond I, II, III	Beta blockers
V	Macrocyclic Antibiotic	Chirobiotic V, T, R, TAG; vancomycin	Polar compounds such as underivatized amino acids
VI	Chiral Crown-Ether	ChiroSil RCA(+); SCA(−); ChiralHyun-CR-1	Amino acids, amino acid esters, amino alcohols
VII	Donor-Acceptor Phases	Whelk-O 1, ULMO, Sumichiral 2500, Sumichiral OA 4900	Amides, epoxides, esters, ureas, carbamates, ethers, aziridines, phosphonates, aldehydes,ketones, carboxylic acids, alcohols
VIII	Chiral Ion-Exchangers	Chiralpak QN-AX; Chiralpak QD-AX	Chiral carboxylic, sulfonic, phosphonic, and phosphoric acids
IX	Chiral Ligand-Exchange	Chiralpak MA+, Nucleosil Chiral-1	Amino acids

Table 3.
Application of nine major types of CSP and their commercial CSP [72].

4.1 Type I polysaccharide-derived CSPs in HPLC

Polysaccharide-derived CSPs are widely used in enantio-separation of a large number of chiral compounds [71]. The development of polysaccharide-derived CSPs has continued for about three decades. It can be roughly divided into two stages (i) the coated CSPs stage, and (ii) the immobilized CSPs stage.

4.1.1 Coated polysaccharide-derived CSPs

Polysaccharide selectors have been used for enantioselective liquid chromatography technique for a long time. In 1973, a polymeric selector (without supporting matrix) was introduced by Hesse and Hagel named as microcrystalline cellulose triacetate (MCTA) used for enantioselective liquid chromatography. MCTA are widely applied in enantio-recognition and preparative separations due to its ideal loading capacities. However, this material has major disadvantages including poor pressure stability, slow separations, and low chromatographic efficiency. To overcome the mechanical stability problem of MCTA, a solution was found by Okamoto and co-workers in 1984, in which the surface of macro-porous silica beads (100 or 400 nm pore size) was coated by the cellulose derivatives at about 20 wt%. Thanks to this coating, the mechanical stability of this material was remarkably improved resulting in better efficiencies qualified for HPLC enantiomer separations. Such coated polysaccharide-based CSPs were the highest level of polymeric selector developments for several decades [71].

Nowadays, about 200 kinds of polysaccharide derivatives were introduced by using different polysaccharides which includes cellulose, amylose, chitin, chitosan, galactosamine, curdlan, dextran, xylan, and inulin [72] (**Figure 13**).

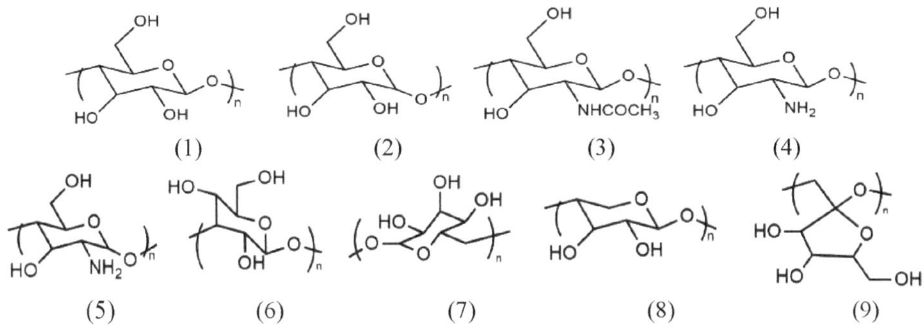

Figure 13.
Structures of the various kinds of polysaccharides: (1) cellulose; (2) amylose; (3) chitin; (4) chitosan; (5) Galatosamine; (6) Curdlan; (7) dextran; (8) Xylan; (9) inulin.

These materials have been coated on a surface of macro-porous silica gel to create CSPs and followed by the evaluation of chiral recognitions on HPLC.

Each of coated polysaccharide-based CSPs exhibit the different enantioselectivity and elution order of the various enantiomers due to the structural differences of CSPs including sugar units, linkage position, and linkage type. In particular, the derivatives of cellulose and amylose usually perform higher recognition abilities than the others, though this property also depends on the structure of a specific racemate. The most useful and successful derivatives of cellulose and amylose are triesters and tricarbamate. It has been claimed by Aboul-Enein and Ali that for the resolution of about 500 test racemates, about 80% of them have been successfully resolved on only two kinds of polysaccharide derivative-based CSPs (cellulose and amylose tris (3,5-diphenylcarbamate) CSPs) [72]. More specifically, three famous commercially available CSPs, CHIRALCEL OD, OJ, and CHIRALPAK AD, have fully or partially resolved 70% racemates among over 100 racemates tested [71].

A current strategy introduced by Snyder for chiral separation method development includes trial-and-error experiments of various polysaccharide-type CSPs under multiple respective mobile-phase conditions using fully automated column- and solvent-switching. Nowadays, more and more studies have focused on developing a more efficient screening procedure to enhance the chance for success and shorten the experiment time: the most favorable CSP in the normal phase mode is

Chiralpak AD– – – > Chiralcel OD– – – – > Chiralcel OJ

The separation efficiency of column should be tested before conducting screening experiment if serial instead of parallel screening is utilized. The application of coated polysaccharide-derived CSPs has been reviewed in **Table 4** including the names of CSPs and their most frequent applications.

4.1.2 Immobilized polysaccharide-derived CSPs

The coated CSPs are formed by coating the polysaccharide derivatives onto surface of silica gel. Due to the weak linkages and interactions between the polysaccharide derivatives (chiral selector) and silicagel (substrate), a number of organic solvents including chloroform, dichloromethane, tetrahydrofuran and ethyl acetate which can dissolve or swell the chiral selector are not allowed

Trade name	Chemical name	Applications
Cellulose CSPs		
Chiralcel OB	Cellulose trisbenzoate	Small aliphatic and aromatic compounds
Chiralcel OJ	Cellulose tris(4-methyl benzoate)	Aryl methyl esters, aryl methoxy esters
Chiralcel OC	Cellulose trisphenylcarbamate	Cyclopentanones
Chiralcel OD	Cellulose tris(3,5-dimethylphenylcarbamate)	Alkaloids, amines, β -adrenergic blockers
Chiralcel OD-H[b]	Cellulose tris(3,5-dimethylphenylcarbamate)	Alkaloids, amines, β -adrenergic blockers
Chiralcel OD-R[c]	Cellulose tris(3,5-dimethylphenylcarbamate)	Alkaloids, amines, β -adrenergic blockers
Chiralcel OD-RH[d]	Cellulose tris(3,5-dimethylphenylcarbamate)	Alkaloids, amines, β -adrenergic blockers
Chiralcel OF	Cellulose tris(4-chlorophenylcarbamate)	β -Lactams, dihydroxypryidines, alkaloids
Chiralcel OG	Cellulose tris(4-methylphenylcarbamate)	β -Lactams, alkaloids
Chiralcel OA	Cellulose triacetate on silica gel	Small aliphatie compounds
Amylose CSPs		
Chiralpak AD	Amylose tris(3,5-dimethylphenylcarbamate)	Alkaloids, tropines, amines, β -adrenergic blockers
Chiralpak AD-R[a]	Amylose tris(3,5-dimethylphenylcarbamate)	Alkaloids, tropines, amines, β -adrenergic blockers
Chiralpak AD-RH[b]	Amylose tris(3,5-dimethylphenylcarbamate)	Alkaloids, tropines, amines, β -adrenergic blockers
Chiralpak AD-H	Amylose tris-(3,5-dimethylphenylcarbamate)	Alkaloids
Chiralpak AR	Amylose tris(R)-1-phenylethylcarbamate	Alkaloids, tropines, amines
Chiralpak AS	Amylose tris(S)-1-methylphenylcarbamate	Alkaloids, tropines, amines

[a]*Columns supplied by Daicel Chemical Industries, Tokyo, Japan, Dimension are column size 25 cm X 0,46 cm, particle size 10 μm, except as noted.*
[b]*Column size 25 cm X 0,46 cm, particle size 5 μm.*
[c]*Column size 15 cm X 0,46 cm, particle size 10 μm.*
[d]*Column size 15 cm X 46 cm, particle size 5 μm.*

Table 4.
Various polysaccharide-based commercial CSP [72, 90, 91, 93].

using as mobile phase components. Besides, the mixtures of alkanes (n-pentane, n-hexane or n-heptane) and alcohols (2-propanol (IPA), ethanol or methanol (n-pentane, n-hexane or n-heptane)) and alcohols (2-propanol (IPA), ethanol or methanol) are favorable mobile phase solvents used in normal phase mode. The addition of "prohibited solvents" may lead to better separation and greater solubility of racemic analytes than the standard solvent combinations so the performance of separation method is improved. Besides, the better solubility of racemic analytes also facilitates preparative separations. In addition, the spectroscopic techniques used for chiral recognition should ideally be in the "prohibited solvents". As a result, chemical immobilization of polysaccharide derivatives becomes an interesting research topic to overcome the drawbacks of the coated

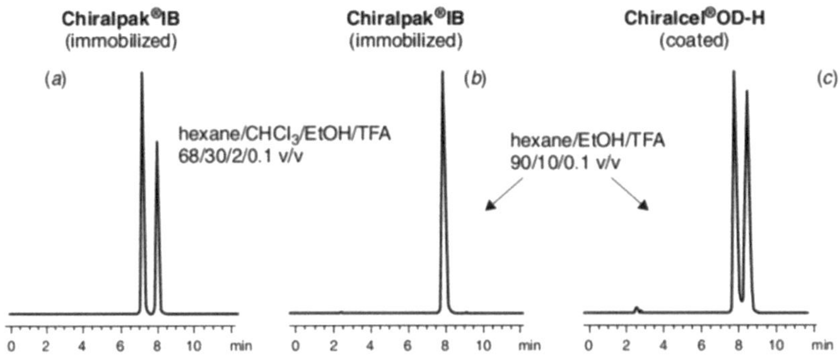

Figure 14.
Effect of column type (immobilized vs. coated polysaccharide-based CSP) and mobile phase on enantioselectivity. Enantiomer separations of N-benzyloxycarbonyl-phenylalanine (a–c) with Chiralpak IB (immobilized) and Chiralcel OD (coated). Flowrate: 1 mL/min; temperature: 25°C; UV detection at 230 nm. Note that the mobile phases used in (a) is forbidden mobile phases for the coated version Chiralcel OD [71].

CSPs (see in **Figure 14**). In 2005, Daicel and Chiral Technologies introduced a set of three immobilized polysaccharide CSPs as [71]:

- Chiralpak®IA (immobilized version of Chiralpak AD)
- Chiralpak®IB (immobilized Chiralcel OD) [61]
- Chiralpak®IC based on the cellulose tris (3,5-dichlorophenylcarbamate) selector that is not available in coated form.

Finally, it should be noted that polysaccharide CSPs have now also been established as the first-choice of chiral phases for enantiomer separation.

5. Analytical methods of chiral alkaloids in medicinal plants

Analytical applications, including CSPs, separation conditions and analyte, are summarized in **Table 5**. A review focused mainly on the latest examples of chiral alkaloids separations on CSPs for efficient analyses was prepared.

Novel column materials also improved enantiomer separation of tropane alkaloids. Separation of (R, S)-hyoscyamine was achieved using chiral stationary phase with immobilizing α-1-acid glycoprotein (Chiral AGP®) [83], alternatively, enantiomer separation of atropine could be achieved by a chirobiotic V column packed with vancomycin as chiral selector [84]. Anisodamine, the 6β-hydroxyl derivative of (S)-hyoscyamine was separated from its synthetic enantiomer and diastereomers by a Chiralpak AD-H column as chiral stationary phase, which uses an amylose derivative as chiral selector [85]. Satropane, 3α-paramethylbenzenesulfonyloxy-6β-acetoxy-tropane, could be resolved in 3S, 6S-isomer named lesatropane and 3R, 6R-isomer named desatropane. Lesatropane as a novel muscarinic agonist is being under preclinical development in China as a single enantiomer drug for the treatment of primary glaucoma. The separation of lesatropane from desatropane was conducted by both Chiralpak AD-H and Chiralpak AS-RH column [86].

Chiral separation of isoquinoline alkaloid has also achieved by using chiral stationary phase. For example, the determination of tetrahydropalmatine (THP) was performed by using a Chiralcel OJ column with quantification by UV at

DOI: http://dx.doi.org/10.5772/intechopen.96009

Columns	Analyte	Plant	Separation conditions	Ref.
CHIRALPAK AS-H column	(12S,22S)-Dihydroxyisoechinulin A (2) and (12R/S)-Neoechinulin A	Cannabis sativ L	Hexane/isopropanol/diethylamine (4:1:0.05)	[73]
CHIRALPAK AD-H column	Mucroniferanine A	Corydalis mucronifera	n-hexane–2-propanol (70:30)	[74]
	(±)-homocrepidine A	Dendrobium crepidatum	n-hexane/2-propanol (95:5)	[75]
	(+)-(3R,6R)- and (−)-(3S,6S)-3α,6β-tropanediol	Erythroxylaceae species	n-hexane and 2-propanol (9:1) with 0.1% of diethylamine	[76]
Chirobiotic V, Chiralpak-AY3 column	(−) and (+) hyoscyamine	Solanaceaes seeds	Ethanol, 0.1% DEA	[77]
Chirex 3019 chiral column	S-(−)-canadine and R-(+)-canadine	*Hydrastis Canadensis* L.	Hexan:DCE:EtOH:TFA (75:40:7:0.1)	[78]
CHIRALPAK AGP column	(R)-nicotine; (S)-nicotine; anabasine, and anatabine	Tobacco	NH_4OH- methanol (90:10)	[79]
A Phenomenex Lux Cellulose-2 chiral column	(+)- and (−)-5-hydroxyl-8-oxyberberine	Coptis chinensis	CH_3CN:H2O (40:60)	[80]
Chiralpak IA column	intermedine and lycopsamine	Symphytum uplandicum	ACN/methanol (80:20) and methanol/methyl-t-butyl ether (90:10)	[81]
Phenomenex-Chirex-3126 column	dihydrocarneamide A and iso-notoamide B	Paecilomyces variotii	MeCN–H_2O (5:95).	[82]

Table 5.
Summary of CSPs, mobile phase compositions, and applications.

230 nm. This developed method was used for determining the pharmacokinetics of THP enantiomers in rats and dogs after oral administration [87]. Another report of isoquinoline alkaloid group is sanguinarine derivatives. Chiral determination of Benzophenanthridine alkaloids from methanol extracts of Hylomecon species was conducted by Chiralcel OD (4.6 x 250 mm) column with mixture of isopropanol-hexane-diethylamine (20/80/0.1, v/v/v) as a mobile phase [88]. The stereochemistry of L-isocorypalmine and the D/L ratio of tetrahydropalmatine, stylopine, and corydaline were established unambiguously by using a chiral Chiralcel OD (4.6 x 250 mm) column; 50% ethanol as mobile phase; wavelength 230 nm [89]. Cularinoids are a group of isoquinoline alkaloids consisting of about 60 members. The HPLC enantiomeric separation of the racemic cularinoid alkaloids N-p-methoxy-1,α-dihydroaristoyagonine and 4′,5’demethoxy-1,α-dihydroaristoyagonine was accomplished using five chiral stationary phases (CSPs), the good enantioselectivity and resolution factor obtained with a polysaccharide-derived CSP (Chiralpak AD) [90].

Indole derivatives are widely used in chiral synthesis, chemical asymmetric catalysis, biological and medicinal chemistry. Recently, the enantiomeric separation of several chiral plant growth regulators and related compounds, such as 3-(3-indolyl)-butyric acid, abscisic acid and structurally related molecules including a variety of substituted tryptophan compounds was reported. Chiral stationary phases such as coated and immobilized were suitable for the separation of indole derivatives; however, the coated CSP possesses a higher resolving power than the immobilized one [91]. Tangutorine, a biogenetically interesting indole alkaloid, was found in the leaves of Nitraria tangutorum in 1999. It was separated from its synthetic enantiomer by chiral stationary phases two polysaccharide-derived CSP (Chiralcel OD and Chiralpak AD) and a network polymer incorporating a bifunctional C_2-symmetric chiral selector (Kromasil CHI-DMB) [92]. The HPLC enantiomeric separation of racemic indole alkaloids tacamonine, 17α-hydroxytacamonine, deethyleburnamonine, and vindeburnol was accomplished using Chiralpak AD and Chiralcel OD as chiral stationary phases [93].

The enantiomers of homocamptothecin (hCPT) derivatives which constitute a promising series of potent anticancer agents targeting DNA topoisomerase I were separated by using the combination of two silica-based normal phase column including Chiralcel OD-H (celluloses tris-3,5-dimethylphenylcarbamateand) Chiralcel OJ (celluloses tris-methylbenzoate) or Chiralpak AD (amyloses tris-3,5-dimethylphenylcarbamate) and Chiralpak AS (amyloses tris-(S)-1-phenylethyl-car-bamate) [94]. A method for the simultaneous determination of eight Cinchona alkaloids (quinine, quinidine, cinchonine, cinchonidine, and their corresponding dihydro analogs) using a novel strong cation-exchange-type chiral stationary phase (cSCX) column in HPLC has been developed and exemplarily applied to impurity profiling of a commercial alkaloid sample [95].

6. Conclusion

Most of alkaloids are chiral compounds and are clinically administered as the racemic mixture, although its enantiomers have been shown to exert different pharmacological activity. The complication of sample matrices and low concentration of chiral alkaloids are major challenges for analytical processes. To improve the performance of analytical procedure, we provide the current state of the art in sample preparation focusing on extraction and purification to remove interferences and enrich analyte concentrations. HPLC using CSPs has demonstrated to be extremely useful, accurate, versatile, its mode is generally the most straightforward and convenient means for chromatographic enantiomer separation. The development of CSPs for HPLC is a continuous and challenger issue covering various types of CSPs.

Acknowledgements

The authors would like to express their hearty gratitude to Can Tho University of Medicine and Pharmacy. We also thank all of our colleagues for their excellent assistance.

Conflict of interest

The authors declare no conflict of interest.

Author details

Ngoc Van Thi Nguyen[1*], Kim Ngan Huynh Nguyen[1], Kien Trung Nguyen[1], Kyeong Ho Kim[2] and Hassan Y. Aboul-Enein[3]

1 Can Tho University of Medicine and Pharmacy, Can Tho City, Vietnam

2 College of Pharmacy, Kangwon National University, Chuncheon-si, Gangwon-do, South Korea

3 Department of Pharmaceutical and Medicinal Chemistry, National Research Center, Egypt

*Address all correspondence to: nguyenthingocvanct@gmail.com; ntnvan@ctump.edu.vn

References

[1] Ekor M. The growing use of herbal medicines: issues relating to adverse reactions and challenges in monitoring safety. Front Pharmacol. 2014 Jan 10;4:177. doi: 10.3389/fphar.2013.00177.

[2] Ravindra Kumar Pandey, Shiv Shankar Shukla and *et al.*, Fingerprinting Analysis and Quality Control Methods of Herbal Medicines, ISBN 9781138036949, Published July 1, 2018 by CRC Press.

[3] Nguyen LA, He H, Pham-Huy C. Chiral drugs: an overview. Int J Biomed Sci. 2006 Jun;2(2):85-100. PMID: 23674971; PMCID: PMC3614593.

[4] Pellati F., and Benvenuti S., Chromatographic and electrophoretic methods for the analysis of phenetylamine alkaloids in Citrus aurantium. J. Chromatogr. A, 1161, 71-88 (2007).

[5] Gurib-Fakim, A., Medicinal plants: traditions of yesterday and drugs of tomorrow. Molecular aspects of medicine, 27, 1-93 (2006).

[6] Wink, M. (2013). Evolution of secondary metabolites in legumes (Fabaceae). South African Journal of Botany, 89. doi: 10.1016/j.sajb.2013.06.006

[7] Evans, W. C., *Pharmacognosy: Part* 5 Pharmacopoeial and related drugs of biological origin, chapter alkaloids, 353-414, 16th edition (2009).

[8] Lesley Braun, Marc Cohen. Herbs and Natural Supplements. 4th ed. Volume 2. Churchill Livingstone Australia November 2014. p 1324.ISBN: 9780729581738.

[9] Amedeo Amedei', Elena Niccolai. Natural Products Analysis: Instrumentation, Methods, and Applications. Chapter 4: Plant and Marine Sources: Biological Activity of Natural Products and Therapeutic Use. p 53. 2014. DOI:10.1002/9781118876015.

[10] Gotti R., Capillary electrophoresis of phytochemical substances in herbal drugs and medicinal plants. J. Pharm. Biomed. Anal. 55, 775-801 (2011).

[11] Karen W. Phinney, Toshihide Ihara, Lane C. Sander, Determination of ephedrine alkaloid stereoisomers in dietary supplements by capillary electrophoresis. J. Chromatogr. A., 1077, 90-97 (2005).

[12] McCooeye, M., Ding, L., Gardner, G. J., Fraser, C. A., Lam, J., Sturgeon, R. E., and Mester, Z., Separation and Quantitation of the Stereoisomers of Ephedra Alkaloids in Natural Health Products Using Flow Injection-Electrospray Ionization-High Field Asymmetric Waveform Ion Mobility Spectrometry-Mass Spectrometry. Anal. Chem., 75, 2538-2542 (2003).

[13] Wang, M., Marriott, P. J., Chan, W-H., Lee, A. W. M., Huie, C. W., Enantiomeric separation and quantification of ephedrine-type alkaloids in herbal materials by comprehensive two-dimensional gas chromatography. J. Chromatogr. A, 1112, 361-368 (2006).

[14] Stohs SJ, Preuss HG, Shara M. A review of the human clinical studies involving Citrus aurantium (bitter orange) extract and its primary protoalkaloid p-synephrine. Int J Med Sci. 2012;9(7):527-538.

[15] Stohs SJ, Preuss HG, Shara M. A review of the receptor-binding properties of p-synephrine as related to its pharmacological effects. Oxid Med Cell Longev. 2011;2011:482973.

[16] Kohnen-Johannsen KL, Kayser O. Tropane Alkaloids: Chemistry, Pharmacology, Biosynthesis and Production. Molecules. 2019;24(4):796. 2019 Feb 22.

[17] Du N, Liu Y, Zhang X, Wang J, Zhao J, He J, Zhou H, Mei L, Liang X. Discovery of new muscarinic acetylcholine receptor antagonists from Scopolia tangutica. Sci Rep. 2017 Apr 7;7:46067.

[18] D. Cardoso, R.T. Pennington, *et al.*, Reconstructing the deep-branching relationships of the papilionoid legumes, South African Journal of Botany. Volume 89. 2013. pp 58-75. https://doi.org/10.1016/j.sajb.2013.05.001.

[19] Liang N, Kong DZ, Nikolova D, Jakobsen JC, Gluud C, Liu JP. Radix Sophorae flavescentis versus antiviral drugs for chronic hepatitis B. Cochrane Database Syst Rev. 2018 Aug 23;2018(8):CD013106. doi: 10.1002/14651858.

[20] Zheng K, Li C, Shan X, Liu H, Fan W, Wang Z. A study on isolation of chemical constituents from Sophora flavescens Ait. and their anti-glioma effects. Afr J Tradit Complement Altern Med. 2013 Nov 2;11(1):156-160. doi: 10.4314/ajtcam.v11i1.24.

[21] Xiao W, Shen G, Zhuang X, Ran X, Zhu M, Li H. Characterization of human metabolism and disposition of levo-tetrahydropalmatine: Qualitative and quantitative determination of oxidative and conjugated metabolites. J Pharm Biomed Anal. 2016 Sep 5;128:371-381.

[22] Wang JB, Mantsch JR. L-tetrahydropalamatine: a potential new medication for the treatment of cocaine addiction. Future Med Chem. 2012 Feb;4(2):177-186. doi: 10.4155/fmc.11.166.

[23] Wu H, Waldbauer K, Tang L, Xie L, McKinnon R, Zehl M, Yang H, Xu H, Kopp B. Influence of vinegar and wine processing on the alkaloid content and composition of the traditional Chinese medicine Corydalis Rhizoma (Yanhusuo). Molecules. 2014 Aug 4;19(8):11487-11504.

[24] Heinrich M. Galanthamine from Galanthus and other Amaryllidaceae--chemistry and biology based on traditional use. Alkaloids Chem Biol. 2010;68:157-165. doi: 10.1016/s1099-4831(10)06804-5.

[25] Smith SW. Chiral toxicology: it's the same thing...only different. Toxicol Sci. 2009 Jul;110(1):4-30. doi: 10.1093/toxsci/kfp097. Epub 2009 May 4.

[26] Carlin MG, Dean JR, Ames JM. Opium Alkaloids in Harvested and Thermally Processed Poppy Seeds. Front Chem. 2020 Aug 27;8:737.

[27] Hindson, B. J., Francis, P. S., Purcell, S. D., Barnett, N. W., Determination of opiate alkaloids in process liquors using capillary electrophoresis. J. Pharm. Biomed. Anal., 43, 1164-1168 (2007).

[28] Molyneux RJ, Gardner DL, Colegate SM, Edgar JA. Pyrrolizidine alkaloid toxicity in livestock: a paradigm for human poisoning? Food Addit Contam Part A Chem Anal Control Expo Risk Assess. 2011 Mar;28(3):293-307.

[29] Li N, Xia Q, Ruan J, Fu PP, Lin G. Hepatotoxicity and tumorigenicity induced by metabolic activation of pyrrolizidine alkaloids in herbs. Curr Drug Metab. 2011 Nov;12(9):823-834.

[30] Chen T, Mei N, Fu PP. Genotoxicity of pyrrolizidine alkaloids. J Appl Toxicol. 2010 Apr;30(3):183-96. doi: 10.1002/jat.1504. PMID: 20112250; PMCID: PMC6376482.

[31] Hamid HA, Ramli AN, Yusoff MM. Indole Alkaloids from Plants as Potential Leads for Antidepressant Drugs: A Mini Review. Front Pharmacol. 2017 Feb 28;8:96. doi: 10.3389/fphar.2017.00096.

[32] Almagro L, Fernández-Pérez F, Pedreño MA. Indole alkaloids from Catharanthus roseus: bioproduction and their effect on human health. Molecules. 2015 Feb 12;20(2):2973-3000.

[33] Patnala, S., and Kanfer, I., HPLC Analysis of Mesembrine-Type Alkaloids in Sceletium Plant Material Used as an African Traditional Medicine. J. Pharm. Pharmaceut. Sci., 13(4), 558-570 (2010).

[34] Fattorusso, E., and Taglialatela-Scafati, O., Modern alkaloids, Chapter 1: Bioactive Alkaloids: Structure and Biology (2008).

[35] Verma, A., Laakso, I., Seppänen-Laakso, T., Huhtikangas, A., and Riekkola, M.-L., A Simplified Procedure for Indole Alkaloid Extraction from Catharanthus roseus Combined with a Semi-synthetic Production Process for Vinblastine. Molecules, 12, 1307-1315 (2007).

[36] Abdel-Salam, O.M.E., Hamdy, S.M., Seadawy, S.A.M. *et al*. Effect of piracetam, vincamine, vinpocetine, and donepezil on oxidative stress and neurodegeneration induced by aluminum chloride in rats. Comp Clin Pathol 25, 305-318 (2016). https://doi.org/10.1007/s00580-015-2182-0.

[37] Gao, Y., Xu, Y., Han, B., Li, J., and Xiang, Q., Sensitive determination of verticine and verticinone inBulbus Fritillariaeby ionic liquid assisted capillary electrophoresis–electrochemiluminescence system. Talanta, 80, 448-453 (2009).

[38] Liu XY, Wang FP. Recent Advances in the Synthesis of Stemona Alkaloids. Nat Prod Commun. 2015 Jun;10(6):1093-1102.

[39] Y. Wen, L. Chen, J. Li, D. Liu, L. Chen, Recent advances in solid-phase sorbents for sample preparation prior to chromatographic analysis, Trends Anal. Chem. 59 (2014) 26-41.

[40] Shang X, Guo X, Li B, Pan H, Zhang J, Zhang Y, Miao X. Microwave-assisted extraction of three bioactive alkaloids from Peganum harmala L. and their acaricidal activity against Psoroptes cuniculi in vitro. J Ethnopharmacol. 2016 Nov 4;192:350-361. doi: 10.1016/j.jep.2016.07.057.

[41] Esclapez, M.D., García-Pérez, J.V., Mulet, A. *et al.* Ultrasound-Assisted Extraction of Natural Products. Food Eng Rev 3, 108 (2011). https://doi.org/10.1007/s12393-011-9036-6.

[42] Vinatoru M. An overview of the ultrasonically assisted extraction of bioactive principles from herbs. Ultrason Sonochem. 2001 Jul;8(3):303-313. doi: 10.1016/s1350-4177(01)00071-2.

[43] Samar Al Jitan, Saeed A. Alkhoori, Lina F. Yousef. Chapter 13 - Phenolic Acids From Plants: Extraction and Application to Human Health. Studies in Natural Products Chemistry, Elsevier,Volume 58, 2018, pages 389-417, https://doi.org/10.1016/B978-0-444-64056-7.00013-1.

[44] Dey P, Kundu A, Kumar A, Gupta M, Lee BM, Bhakta T, Dash S, Kim HS. Analysis of alkaloids (indole alkaloids, isoquinoline alkaloids, tropane alkaloids). Recent Advances in Natural Products Analysis. 2020:505-567. doi: 10.1016/B978-0-12-816455-6.00015-9.

[45] Linqiu Li, Mingyuan Huang, Junli Shao, Bokun Lin, Qing Shen. Rapid determination of alkaloids in Macleaya cordata using ionic liquid

extraction followed by multiple reaction monitoring UPLC–MS/MS analysis. Journal of Pharmaceutical and Biomedical Analysis. Volume 135. 2017. Pages 61-66.

[46] Atlabachew M, Chandravanshi BS, Redi-Abshiro M. Preparative HPLC for large scale isolation, and salting-out assisted liquid-liquid extraction based method for HPLC-DAD determination of khat (Catha edulis Forsk) alkaloids. Chem Cent J. 2017 Oct 17;11(1):107. doi: 10.1186/s13065-017-0337-6.

[47] Nowak J, Woźniakiewicz M, Klepacki P, Sowa A, Kościelniak P. Identification and determination of ergot alkaloids in Morning Glory cultivars. Anal Bioanal Chem. 2016 May;408(12):3093-3102. doi: 10.1007/s00216-016-9322-5.

[48] Soib HH, Ismail HF, Husin F, Abu Bakar MH, Yaakob H, Sarmidi MR. Bioassay-Guided Different Extraction Techniques of Carica papaya (Linn.) Leaves on In Vitro Wound-Healing Activities. Molecules. 2020 Jan 24;25(3):517. doi: 10.3390/molecules25030517.

[49] Delazar A, Nahar L, Hamedeyazdan S, Sarker SD. Microwave-assisted extraction in natural products isolation. Methods Mol Biol. 2012;864:89-115. doi: 10.1007/978-1-61779-624-1_5.

[50] Chan CH, Yusoff R, Ngoh GC, Kung FW. Microwave-assisted extractions of active ingredients from plants. J Chromatogr A. 2011 Sep 16;1218(37):6213-6225. doi: 10.1016/j.chroma.2011.07.040. Epub 2011 Jul 23.

[51] Luiz Carlos Klein-Júnior, Yvan Vander Heyden, Amélia Teresinha Henriques. Enlarging the bottleneck in the analysis of alkaloids: A review on sample preparation in herbal matrices. TrAC Trends in Analytical Chemistry, Volume 80, 2016, Pages 66-82.

[52] Kopp T, Abdel-Tawab M, Mizaikoff B. Extracting and Analyzing Pyrrolizidine Alkaloids in Medicinal Plants: A Review. Toxins (Basel). 2020 May 13;12(5):320. doi: 10.3390/toxins12050320.

[53] Shang X, Guo X, Li B, Pan H, Zhang J, Zhang Y, Miao X. Microwave-assisted extraction of three bioactive alkaloids from Peganum harmala L. and their acaricidal activity against Psoroptes cuniculi in vitro. J Ethnopharmacol. 2016 Nov 4;192:350-361. doi: 10.1016/j.jep.2016.07.057.

[54] Dao-Tao Xie, Ya-Qin Wang, Yun Kang, Qiu-Fen Hu, Na-Ya Su, Jian-Ming Huang, Chun-Tao Che, Ji-Xian Guo. Microwave-assisted extraction of bioactive alkaloids from Stephania sinica. Separation and Purification Technology. Volume 130. 2014. Pages 173-181.

[55] Xiong W, Chen X, Lv G, Hu D, Zhao J, Li S. Optimization of microwave-assisted extraction of bioactive alkaloids from lotus plumule using response surface methodology. J Pharm Anal. 2016 Dec;6(6):382-388.

[56] Mao Z, Di X, Zhang J, Wang X, Liu Y, Di X. Rapid and cost-effective method for the simultaneous quantification of seven alkaloids in Corydalis decumbens by microwave-assisted extraction and capillary electrophoresis. J Sep Sci. 2017 Jul;40(14):3008-3014. doi: 10.1002/jssc.201700051.

[57] Wei J, Chen J, Liang X, Guo X. Microwave-assisted extraction in combination with HPLC-UV for quantitative analysis of six bioactive oxoisoaporphine alkaloids in Menispermum dauricum DC. Biomed Chromatogr. 2016 Feb;30(2):241-248. doi: 10.1002/bmc.3541.

[58] Jerry W. King, Fundamentals and Applications of Supercritical Fluid

Extraction in Chromatographic Science, Journal of Chromatographic Science, Volume 27, Issue 7, July 1989, Pages 355-364.

[59] Ernesto Reverchon, Iolanda De Marco. Supercritical fluid extraction and fractionation of natural matter. The Journal of Supercritical Fluids,Volume 38, Issue 2, 2006, Pages 146-166. https://doi.org/10.1016/j.supflu.2006.03.020.

[60] Rui P.F.F. da Silva, Teresa A.P. Rocha-Santos, Armando C. Duarte. Supercritical fluid extraction of bioactive compounds. TrAC Trends in Analytical Chemistry. Volume 76, 2016, Pages 40-51. https://doi.org/10.1016/j.trac.2015.11.013.

[61] Miguel Herrero, Alejandro Cifuentes, Elena Ibañez. Sub- and supercritical fluid extraction of functional ingredients from different natural sources: Plants, food-by-products, algae and microalgae: A review. Food Chemistry, Volume 98, Issue 1, 2006, Pages 136-148. https://doi.org/10.1016/j.foodchem.2005.05.058.

[62] Kaufmann B, Christen P. Recent extraction techniques for natural products: microwave-assisted extraction and pressurised solvent extraction. Phytochem Anal. 2002 Mar-Apr;13(2):105-113. doi: 10.1002/pca.631.

[63] Ezel Boyacı, Ángel Rodríguez-Lafuente, Krzysztof Gorynski, Fatemeh Mirnaghi, Érica A. Souza-Silva, Dietmar Hein, Janusz Pawliszyn. Sample preparation with solid phase microextraction and exhaustive extraction approaches: Comparison for challenging cases, Analytica Chimica Acta, Volume 873, 2015. Pages 14-30. https://doi.org/10.1016/j.aca.2014.12.051.

[64] Thurman, E. M., & Mills, M. S. (1998). Solid-phase extraction: Principles and practice. New York: Wiley. Pp. 1-48.

[65] Teixeira J, Tiritan ME, Pinto MMM, Fernandes C. Chiral Stationary Phases for Liquid Chromatography: Recent Developments. Molecules. 2019 Feb 28;24(5):865. doi: 10.3390/molecules24050865.

[66] Lämmerhofer, M. Chiral recognition by enantioselective liquid chromatography: Mechanisms and modern chiral stationary phases. J. Chromatogr. A 2010, 1217, 814-856, doi:10.1016/j.chroma.2009.10.022.

[67] Fernandes, C.; Tiritan, M.E.; Cravo, S.; Phyo, Y.Z.; Kijjoa, A.; Silva, A.M.S.; Cass, Q.B.; Pinto, M.M.M. Newchiral stationary phases based on xanthone derivatives for liquid chromatography. Chirality 2017, 29, 430-442, doi:10.1002/chir.22706.

[68] Scriba, G.K.E. Chiral recognition in separation science—An update. J. Chromatogr. A 2016, 1467, 56-78, doi:10.1016/jchroma.2016.05.061.

[69] Kalíková, K.; Riesová, M.; Tesařová, E. Recent chiral selectors for separation in HPLC and CE. Cent. Eur. J. Chem. 2011, 10, 450-471, doi:10.2478/s11532-011-0142-3.

[70] Ali I, Suhail M, Asnin L. Chiral separation of quinolones by liquid chromatography and capillary electrophoresis. J Sep Sci. 2017 Jul;40(14):2863-2882. doi: 10.1002/jssc.201700200.

[71] Snyder, L. R., Kirkland, J. J., and Dolan, J. W., Introduction to modern liquid chromatography, 3rd ed., 666-715 (2010).

[72] Aboul-Enein, H. Y., and Ali, I., Chiral Separations by Liquid Chromatography and Related Technologies, 90, chapter 2 (2003).

[73] Xiaoli Yan, Yuefang Zhou, Jiajing Tang, Mei Ji, Hongxiang Lou, Peihong Fan. Diketopiperazine indole alkaloids

from hemp seed. Phytochemistry Letters,Volume 18, 2016, Pages 77-82. https://doi.org/10.1016/j.phytol.2016.09.001

[74] Zhang J, Zhang QY, Tu PF, Xu FC, Liang H. Mucroniferanines A-G, Isoquinoline Alkaloids from Corydalis mucronifera. J Nat Prod. 2018 Feb 23;81(2):364-370. doi: 10.1021/acs.jnatprod.7b00847.

[75] Hu Y, Zhang C, Zhao X, Wang Y, Feng D, Zhang M, Xie H. (±)-Homocrepidine A, a Pair of Anti-inflammatory Enantiomeric Octahydroindolizine Alkaloid Dimers from Dendrobium crepidatum. J Nat Prod. 2016 Jan 22;79(1):252-256. doi: 10.1021/acs.jnatprod.5b00801.

[76] Muñoz, Marcelo & González, Natalia & Joseph-Nathan, Pedro. (2016). Enantiomeric High-performance liquid chromatography resolution and absolute configuration of 6β-benzoyloxy-3α-tropanol. Journal of Separation Science. 39. 10.1002/jssc.201600061.

[77] Jesús Marín-Sáez, Roberto Romero-González, Antonia Garrido Frenich, Enantiomeric determination and evaluation of the racemization process of atropine in Solanaceae seeds and contaminated samples by high performance liquid chromatography-tandem mass spectrometry. Journal of Chromatography A, Volume 1474, 2016, Pages 79-84.

[78] Francisco García Sánchez, Aurora Navas Díaz, Ignacio Medina Lama, Alfonso Aguilar & Manuel Algarra (2014). HPLC enantioseparation of the alkaloid canadine and determination of enantiomeric purity with chiral/photometric and achiral/polarimetric detection, Journal of Liquid Chromatography & Related Technologies, 37:1, 26-38, DOI: 10.1080/10826076.2012.733996.

[79] Huihua, Ji & Wu, Ying & Fannin, Franklin & Bush, Lowell. (2019). Determination of tobacco alkaloid enantiomers using reversed phase UPLC/MS/MS. Heliyon. 5. e01719. 10.1016/j.heliyon.2019.e01719.

[80] Wang, Lei & Zhang, Sheng-Yuan & Chen, Liang & Huang, Xiao-Jun & Zhang, Qing-Wen & Jiang, ren-wang & Yao, Fen & Ye, Wen-Cai. (2014). New enantiomeric isoquinoline alkaloids from Coptis chinensis. Phytochemistry Letters. 7. 89-92. 10.1016/j.phytol.2013.10.007.

[81] Pawar, Rahul & Grundel, Erich & Mazzola, Eugene & White, Kevin & Krynitsky, Alexander & Rader, Jeanne. (2010). Chiral stationary phases for separation of intermedine and lycopsamine enantiomers from Symphytum uplandicum. Journal of separation science. 33. 200-5. 10.1002/jssc.200900611.

[82] Peng Zhang, Xiao-Ming Li, Jia-Ning Wang, Xin Li, Bin-Gui Wang. Prenylated indole alkaloids from the marine-derived fungus Paecilomyces variotii. Chinese Chemical Letters, Volume 26, Issue 3, 2015, Pages 313-316.

[83] Soares, R., Singh, A. K., Kedor-Hackmann, E. R. M., and Santoro, M. I. R. M., Determination of Atropine Enantiomers in Ophthalmic Solutions by Liquid Chromatography Using a Chiral AGP® Column. J. AOAC. INT., 92, 1663-1672 (2009).

[84] Aehle, E., and Drager, B., Tropane alkaloid analysis by chromatographic and electrophoretic techniques: An update. J. Chromatogr. B., 878, 1391-1406 (2010).

[85] Yang, L.-M., Xie, Y.-F., Chen, H.-Z., and Lu, Y., Diastereomeric and enantiomeric high-performance liquid chromatographic separation of synthetic anisodamine. J. Pharm. Biomed. Anal., 43, 905-909 (2007).

[86] Yang, L.M., Xie, Y. F., Gu, Z. H., Wang, A. L., Chen, H. Z., and Lu, Y., Development and Validation of a Reversed-Phase HPLC Method for Determination of Lesatropane and Enantiomeric Impurity. Chirality, 23, 581-584 (2011).

[87] Hong, Z., Fan, G., Chai, Y., Yin, X., and Wu, Y., Stereoselective Pharmacokinetics of Tetrahydropalmatine after Oral Administration of (−)-Enantiomer and the Racemate. Chirality, 17, 293-296 (2005).

[88] Kang, J. S., Pham, H. L., Lim, H. M., Kim, Y. H., and Blaschke, G., Achiral and chiral determination of Benzophenanthridine alkaloids from methanol extracts of Hylomecon species by HPLC. Arch. Pharm. Res., 26, 114-119 (2003).

[89] Ma, Z-Z., Xu, W., Jensen, N. H., Roth, B. L., Liu-Chen, L-Y., and Lee, D. Y. W., Isoquinoline Alkaloids Isolated from Corydalis yanhusuo and Their Binding Affinities at the Dopamine D1 Receptor. Molecules, 13, 2303-2312 (2008).

[90] Caccamese, S., Scivoli, G., Bianca, S., L'opez-Romero, J. M., and Ortiz-L'opez, F. J., Chiral high-performance liquid chromatographic separation and circular dichroism spectra of the enantiomers of cytotoxic aristocularine alkaloids. J. Chromatogr. A, 1129, 140-144 (2006).

[91] Zhao, L., Ming, Y., Zhang, H. L., and Jiang, S., Enantioselective Separation of Indole Derivatives by Liquid Chromatography Using Immobilized Cellulose (3,5-Dimethylphenylcarbamate) Chiral Stationary Phase. J. Anal. Chem. 64, 795-805 (2009).

[92] Putkonen, T., Tolvanen, A., Jokela, R., Caccamese, S., and Parrinello, N., Total synthesis of (±)-tangutorine and chiral HPLC separation of enantiomers. Tetrahedron, 59, 8589-8595 (2003).

[93] Caccamese, S., Principato, G., Jokela, R., Tolvanen, A., and Belle, D., Chiral HPLC Separation and CD Spectra of the Enantiomers of the Alkaloid Tacamonine and Related Compounds. Chirality, 13, 691-693 (2001).

[94] Goossens, J. F., Foulon, C., Bailly, C., Bigg, D. C. H., Bonte, J. P., and Vaccher, C., Chiral Resolution of Enantiomers of Homocamptothecin Derivatives, Antitumor Topoisomerase I Inhibitors, Using High Performance Liquid Chromatography on Polysaccharide-Based Chiral Stationary Phases. Chromatographia, 59, 305-313 (2004).

[95] Jasiewicz, B., and Wyrzykiewicz, E., Mass Spectrometry of Bis-quinolizidine Alkaloids: Differentiation of Stereoisomers and Metamers Using ESI and FAB Mass Spectrometry. Spectroscopy Letters, 42, 49-57 (2009).

Chapter 5

Role of Click Chemistry in Organic Synthesis

Ayushi Sethiya, Nusrat Sahiba and Shikha Agarwal

Abstract

Click chemistry involves highly efficient organic reactions of two or more highly functionalized chemical entities under eco-benign conditions for the synthesis of different heterocycles. Several organic reactions such as nucleophilic ring-opening reactions, cyclo-additions, nucleophilic addition reactions, thiol-ene reactions, Diels Alder reactions, etc. are included in click reactions. These reactions have very important features *i.e.* high functional group tolerance, formation of a single product, high atom economy, high yielding, no need for column purification, etc. It also possesses several applications in drug discovery, supramolecular chemistry, material science, nanotechnology, etc. Being highly significant and valuable, we have elaborated on several aspects of click reactions in organic synthesis in this chapter. Recent advancements in the field of organic synthesis using click chemistry approach have been deliberated by citing last five years articles.

Keywords: click chemistry, organic synthesis, eco-benign synthesis, selectivity, atom economy, cyclo-addition

1. Introduction

Presently, researchers are paying considerable attention to devise eco-friendly approaches for organic transformations. There has been a significant hike in interest among the scientists for more environmentally acceptable processes in the chemical industries. Synthetic chemistry has led us to the development of more potent structural analogs of natural products. The high therapeutic efficiency, bioavailability, and pharmacological characteristics of synthetic molecules have increased their use in medicinal chemistry as compared to natural products. Pharmaceutical chemistry encompasses the design, synthesis, and evaluation of compounds. In designing drugs, there is an upsurge demand for eco-benign pathways to accomplish the green aspects of chemistry. Novel green pathways play a vital role in the synthetic chemistry field by eradication of harmful solvents and chemicals or suitable handling of waste materials. The quest for new and proficient approaches for the synthesis of numerous biologically active scaffolds has made click chemistry a promising approach in chemistry. Click chemistry is a fruitful approach for the fabrication of molecules.

Huisgen and co-workers demonstrated a click reaction, Cu(I)-catalyzed azide–alkyne cycloaddition (CuAAC). The advanced use of this reaction and click chemistry was introduced by K. Barry Sharpless in 2001. The term click chemistry not only refers to the reaction which occurs fast but also to those that involve twelve principles of green chemistry *i.e.* selective, easier product isolation, mild reaction condition, high yield, good atom economy, avoid toxic catalyst and solvent, and so on. They encompass reactions that are high yielding, fast, modular, and wide in scope. They are practical and tolerant for a variety of functional groups. Finally, the product isolation is expected to be effortless due to lack of by-products. These vast characteristics make click chemistry a powerful tool that paves a path in several fields of research *viz.* designing of drugs and lead structures [1–4]. Therefore the synergy between these disciplines has given rise to an area of intense research activity. The click chemistry has been such an engrossing topic of research that a lot of review articles have been published so far which explained its applicability in various fields of chemistry like manufacturing and alteration of metal–organic frameworks [5], making devices in bio-sensing system for responsive copper identification [6], designing bio-adhesives for hastening wound closure [7], in virus-related research [8], generation of biosensors [9], proteomics analysis [10], in strategy for indirect ^{18}F-labeling [11], *in vivo* bio-imaging [12], to identify the interaction of curcumin with protein [13], synthesis of polymers and material science [14], for surface modification [15], and so on. In this chapter, the state-of-art modernization with a particular focus on click chemistry assisted synthesis and their uses in various fields of science have been discussed. An attempt has been done to prepare an outline of the importance of click chemistry and its foremost requirement in the research area. It is predicted that this methodical study will pave the way for future opportunities in this direction and design of safer, economical, and eco-friendly pathways.

2. Green aspects of click reactions

According to Sir John Cornforth, a Noble Prize laureate in chemistry in 1975, ideal reaction has been defined as "The ideal chemical process is that which a one-armed operator can perform by pouring the reactants into a bath-tub and collecting the pure product from the drain hole" [16]. Click reactions are designed in such a way that it involves all the twelve principles of green chemistry. Click chemistry includes synthetic methods that are designed to maximize the inclusion of all resources used in the process into the final product. Due to involvement of addition and rearrangement reactions, they have high atom economy. The products are designed with maximum efficacy and minimum cytotoxicity [17]. The green aspects have been depicted in **Figure 1**.

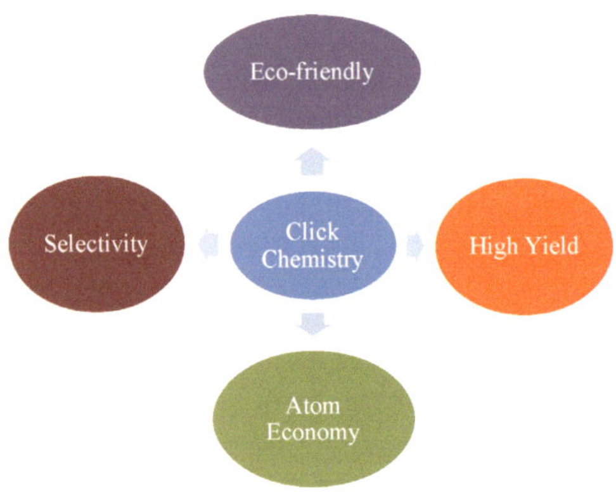

Figure 1.
Green aspects of click chemistry.

3. Role of click reactions in synthetic chemistry

Click chemistry includes a cluster of powerful linking chemical reactions that are easy to perform, have high yields, require no or minimal purification, and are flexible in the unification of different structures without the prerequisite of protection steps. Molecular diversity, modularity, and efficiency are essential in synthetic organic chemistry and expected to be involved in the preparation of several complexes and multi-purpose compounds. In general "Click Chemistry" is a class of biocompatible reactions, to link desired substrates with particular biomolecules. Natural products are produced by joining tiny modular units *via* biosynthesis as well as photosynthesis [18]. Click chemistry provides a route for the synthesis of several heterocyclic scaffolds, amino acids, triazole-fused heterocycles, peptides, and chromophores [19, 20].

3.1 Classification of click reaction

There is no specific classification of click reactions. The chief requisite for "Click Chemistry" is well met by reactions that take place in nature and their mimic in the laboratory is the closest and most desirable to the mind and spirit of most synthetic organic chemists. Usually, four main classifications of click reactions have been identified [21, 22].

- **Cycloadditions**: These refers to 1,3-dipolar cycloadditions reactions and hetero-Diels-Alder cycloadditions.

- **Nucleophilic ring-opening reactions**: This classification belongs to the opening of strained heterocyclic electrophiles, such as epoxides, aziridines, aziridinium ions, cyclic sulfates episulfonium ions, etc.

- **Nucleophilic addition reaction:** It includes the reaction of carbonyl groups like formation of hydrazones, urease, thiourease, oxime ethers, aromatic heterocycles, amides, etc.

- **Additions to carbon–carbon multiple bonds:** It involves epoxidation, dihydroxylation, aziridination, nitrosyl halide addition, sulfenyl halide addition, and certain Michael additions.

Presently, click chemistry inspired synthesis has become the most fascinating approach. Several multi-component reactions have been designed in an eco-friendly manner like aldol condensation followed by Michael addition, Ugi reaction/aldol reaction, Ugi reaction/Huisgen reaction, Ugi Reaction/Diels-Alder reaction, Ugi reaction/Heck reactions, Michael addition/Mannich reaction, etc. [23]

The most famous click reaction is the classical reaction between an azide and an alkyne. Both the substrates do not react under physiological conditions and go through a cycloaddition reaction only at a particular temperature. The uncatalyzed reaction is usually slow and not regio-selective. On the other hand, it was found that the use of electron-deficient terminal alkynes can cause 1,4-regioselectivity to a great extent. These factors limit the use of uncatalyzed Huisgen cycloaddition as an efficient conjugation pathway [24].

3.2 Metal-catalyzed approach for the click synthesis

3.2.1 Synthesis of triazole derivatives

Metals have been used to catalyze several click reactions. The mechanism of metal catalyzed azide alkyne click reaction involves formation of π-alkynyl complex with metal followed by complexation of azide by metal of the π-coordinated triple bond. After cyclization, metallacycle is formed followed by the reductive elimination to afford the relevant 1,2,3-triazole. Several metal like Cu, Ru, Ag, Au, etc. have been employed to accomplish click reactions [25–28]. This section has been divided in two subsections:

3.2.1.1 Metal-catalyzed synthesis of triazole derivatives

Transition metals have been used to catalyze several organic reactions as they provide large surface area and they have vacant d-orbitals due to which they can show variable oxidation state that help in generation of intermediate for organic synthesis [29, 30]. The common process for the click reaction is the transition metal catalyzed synthesis of 1,2,3-triazoles. 1,3-dipolar cyclo-addition of an azide and an alkyne catalyzed by Cu is the most extensively used click-chemistry pathway due to its high selectivity and simplicity [31]. In 2014, Guo and co-workers synthesized β-cyclodextrin derivatives (1) using mono-6-azidocyclodextrin and aromatic aldehydes by Cu^{I}-catalyzed azide–alkyne cyclo-addition. The mono, di, and tri derivatives were synthesized upto 75% yield under mild reaction conditions [32] **(Figure 2).**

Later on, Kumar *et al.* [33] designed a library of new nucleosides **(2 and 3)** having 1,2,3-triazole scaffold at the 2″-position of the sugar nucleus. It was synthesized by 2″-azidouridine using the copper (I)-catalyzed Huisgen–Sharpless–Meldal 1,3-dipolar cyclo-addition reaction **(Figure 3).** The reaction gave 52–82% yield and 1,4-disubstituted 1,2,3-triazoles were obtained.

Tale and co-workers also synthesized 1,2,3-triazoles in excellent yields using (1-(4-methoxybenzyl)-1-*H*-1,2,3-triazol-4-yl)methanol (MBHTM) ligand (1.1 mol%) and $CuSO_4$ (1 mol%) as a catalyst and sodium ascorbate (5 mol%) in DMSO:H_2O(1:3) as a solvent [34]. Shamla and co-workers synthesized coumarin substituted triazole derivatives (4) in good yields using 4-bromomethylcoumarins, terminal alkynes, and sodium azide in the presence of triethylamine and CuI as a catalyst [35] (**Figure 4**).

Yarlagadda *et al*. synthesized N-((l-benzyl-sl*H*-l,2,3-triazol-5-yl) methyl)-4-(6-methoxy benzo[d]thiazol-2-yl)-2-nitrobenzamide derivatives (5) and examined their anti-microbial activity. Among these compounds, compounds **5a, 5 h, 5i** possessed promising activity in comparison to standard drug ciprofloxacin and miconazole (**Figure 5**) [36].

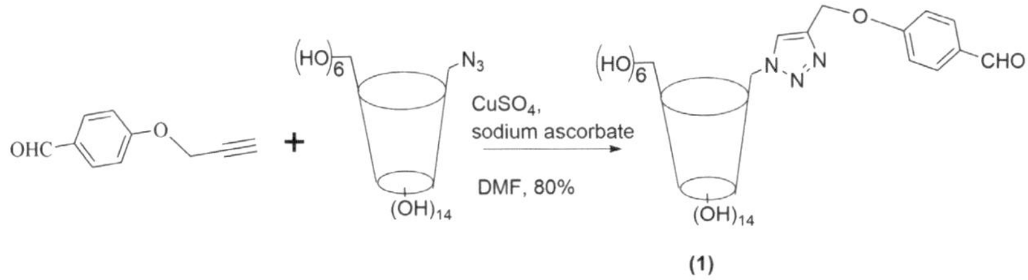

Figure 2.
Synthesis of β-cyclodextrin derivatives using click chemistry approach.

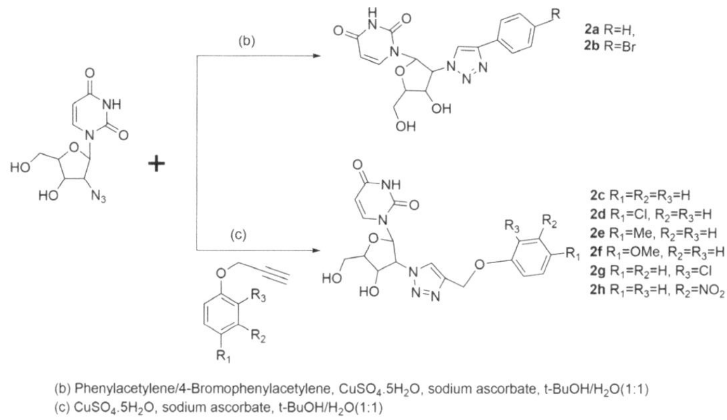

Figure 3.
Synthesis of library of traizole substituted nucleosides.

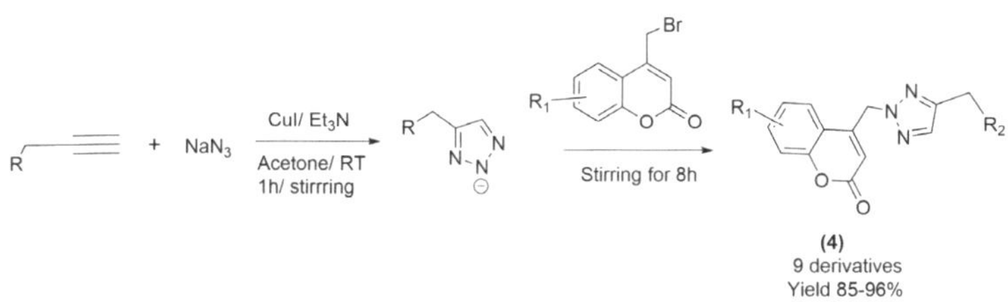

Figure 4.
Synthesis of coumarin substituted triazole derivatives.

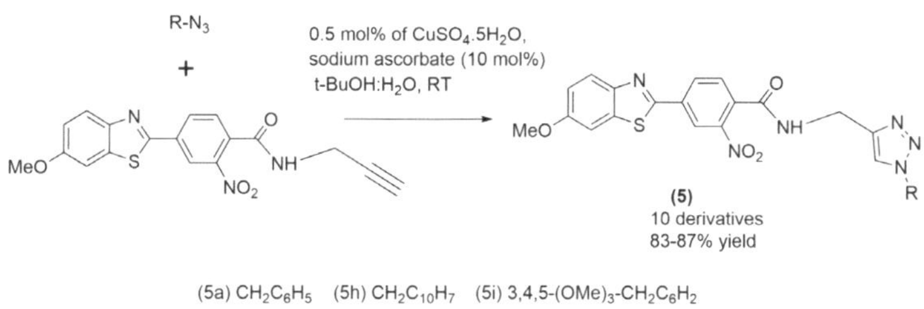

Figure 5.
Synthesis of N-((l-benzyl-lH-l,2,3-triazol-5-yl) methyl)-4-(6-methoxy benzo[d]thiazol-2-yl)-2-nitrobenzamide derivatives.

Anand *et al*. designed Cu(I) catalyzed regio-selective synthesis of iso-indoline-1,3-dione linked 1,4 coumarinyl 1,2,3-triazoles (6) and Ru (II) catalyzed pathway of 1,5 coumarinyl 1,2,3-triazoles in high yields with no need for further purification (**Figure 6**) [37].

Anandhan *et al*. prepared a series of triazole-based macrocyclic amides (7) *via* click chemistry using $CuSO_4$ (5 mol%), sodium ascorbate (10 mol %) in the presence of H_2O/THF (1:1), RT. The synthesized compounds showed good anti-inflammatory activity although at low concentration (50 µg/mL) in comparison to the reference drug prednisolone (**Figure 7**) [38].

Li and co-workers designed triazole derivatives (8 and 9) by click chemistry using $CuSO_4{\bullet}5H_2O$ (0.1 g) and ascorbic acid (0.1 g) in tBuOH/H_2O as a solvent and investigated their applications to synthesize self-assembled membrane against copper corrosion. As per the investigation results, it was found that 2-(1-tosyl-1*H*-1,2,3-triazol-4-yl)-ethanol (TTE) (8) and 2-(1-tosyl-1*H*-1,2,3-triazol-4-yl)-propan-2-ol (TTP) (9) coating on film can sturdily decrease the corrosion caused by copper in

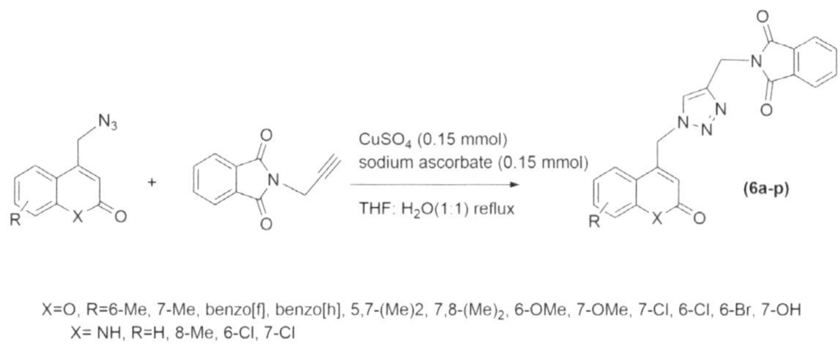

Figure 6.
Synthesis of iso-indoline-1,3-dione linked 1,4 coumarinyl 1,2,3-triazoles derivatives.

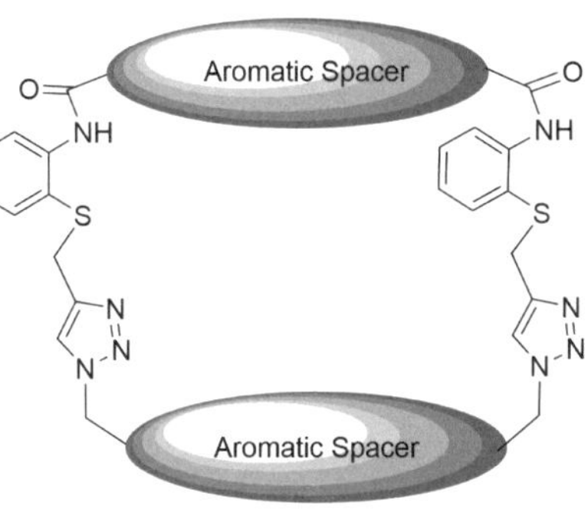

Figure 7.
Triazole based macrocyclic amides.

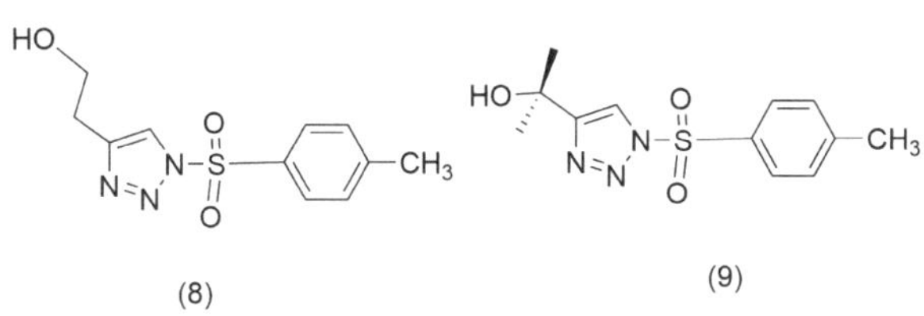

Figure 8.
Derivatives of triazole.

3 wt.% NaCl solution and the inhibition effectiveness of TTP and TTE were 93.1% and 89.4%, respectively (**Figure 8**) [39].

Savanur and co-workers developed facile click chemistry inspired synthesis of triazole ring fused coumarin and quinolinone derivatives using $CuSO_4$ (10 mol%), sodium ascorbate (10 mol%), H_2O:PEG, RT followed by $K2CO_3$/DMF at 50–60 °C and examined their anti-microbial activity. Among the synthesized compounds, compounds **10j, 11 g** and **12f** displayed good anti-bacterial activities. Derivatives **10e** and **10j** were found highly active against yeast strains. Compound **11f** was highly active against filamentous strain *A. niger* and yeast fungi [40] (**Figure 9**).

Yarovaya *et al.* [41] fabricated a conjugate of cytisine with camphor having triazole ring using click chemistry pathway by employing $CuSO_4$•$5H_2O$, sodium ascorbate, t-BuOH/H_2O. The designed molecules were examined for *in vitro* antiviral activity against A/PuertoRico/8/34 influenza virus (H1N1). The compound (13) has highest inhibition activity with IC_{50} = 8 ± 1 μmol (**Figure 10**).

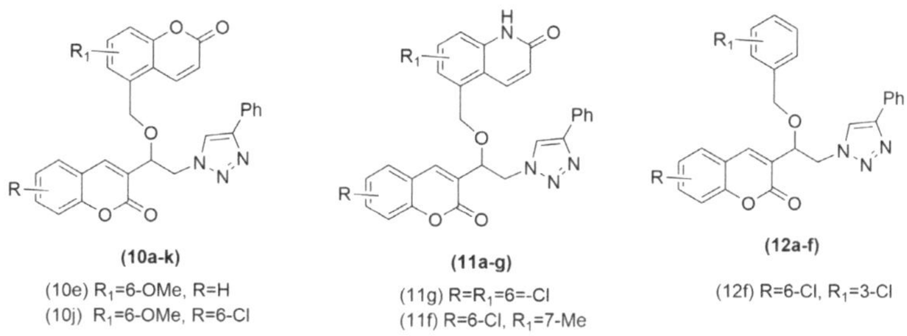

Figure 9.
Triazole ring fused coumarin and quinolinone derivatives.

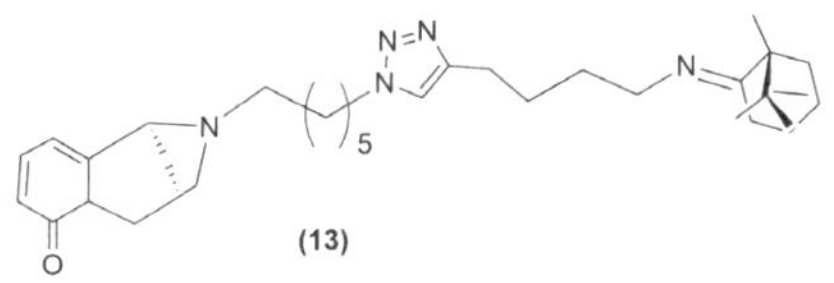

Figure 10.
Cytisine conjugated triazole derivative.

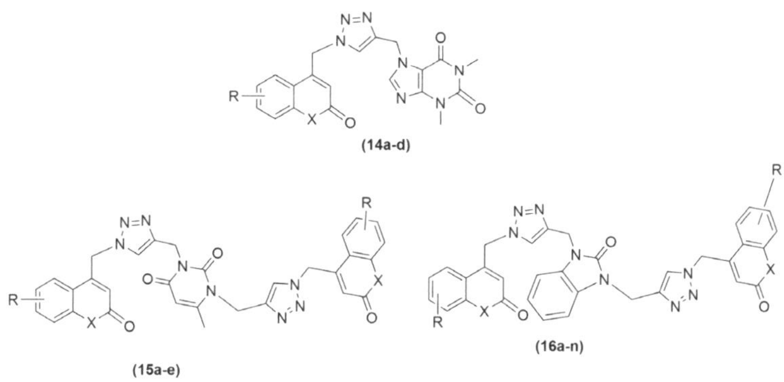

Figure 11.
Methyluracil and theophylline mono-triazole compounds and Bis-triazole compounds.

Khanapurmath *et al*. synthesized various derivatives of triazole by click chemistry approach using $CuSO_4$ and ascorbic acid in H_2O:DMF solvent and assessed them against *Mycobacterium tuberculosis H37Rv*. 6-Methyluracil and theophylline mono-triazole compounds **14(a-d)** and bis-triazole compounds, **15(a-e)** showed reasonable inhibition of *M. tuberculosis H37Rv*, with MIC values in the range of 55.62–115.62 μM. Benzimidazolone bis-triazole derivatives **16(a-n)** inhibited *M. tuberculosis H37Rv* with MIC 2.35–18.34 μM (**Figure 11**) [42].

3.2.1.2 Metal Nano-particle based triazole synthesis

Green synthesis is the fundamental requirement of present synthetic protocol and use of nanoparticles (Nps) is one of the key tackle for organic transformations. NPs are microscopic particles with dimension between 1–100 nm. These are used as catalysts because they provide large surface area, high catalytic activity, nontoxic, heterogeneous nature, etc. In lieu of this, Chetia *et al*. designed copper Nps (nano particles) supported over hydrotalcite and used these Nps (15 mg) to catalyze 1,3 dipolar cycloaddition reaction to form 1,4 disubstituted-1,2,3-triazoles (17)

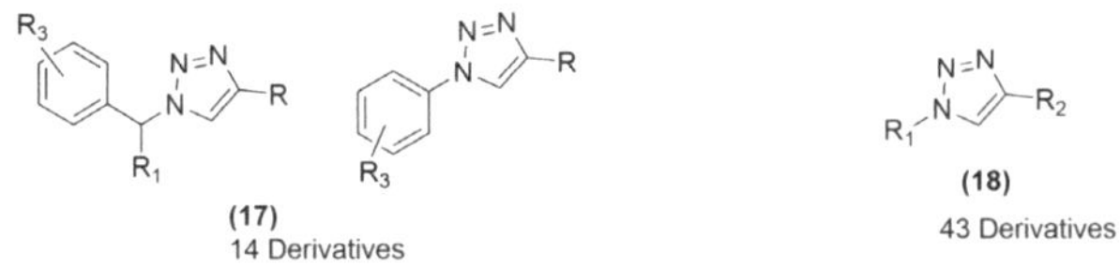

Figure 12.
Triazole derivatives.

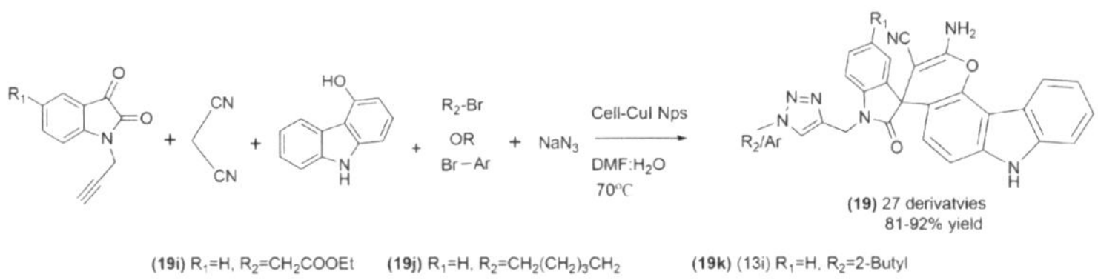

Figure 13.
Spirochromenocarbazole tethered 1,2,3-triazole derivatives.

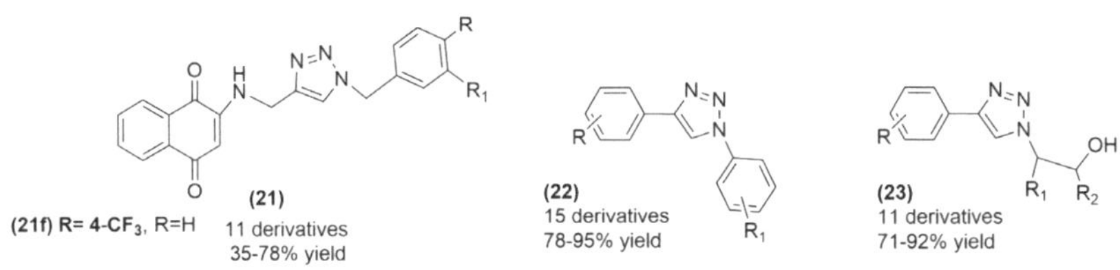

Figure 14.
Structures of different triazole derivatives.

(**Figure 12**) at room temperature using ethylene glycol as a solvent. The catalyst is heterogeneous, easily recyclable, and further reusable [43]. In this context, Poshala and co-workers developed copper Nps (0.1 mmol) using rongalite as a reducing agent and examined their catalytic efficiency in synthesizing triazole (18) derivatives in the presence of β-cyclodextrin (0.02 mmol) [44].

Chavan *et al.* [45] designed a click chemistry assisted MCR strategy for the synthesis of spirochromenocarbazole tethered 1,2,3-triazoles (19) using CuINps supported over cellulose (7 mol%) as a catalyst in the presence of DMF:Water (1:2) (**Figure 13**). The synthesized compounds were screened for anti-cancer activity against MCF-7, HeLa, MDA-MB-231, A-549, PANC-1 and THP-1. Compounds **19i** and **19j** were observed to be the most potent against MCF-7 with IC_{50} = 2.13 μM and 4.80 μM respectively. Compound **19 k** was the most potent one against MDA-MB-231 with (IC_{50} = 3.78 μM). All the products were found to be safe against the human umbilical vein endothelial cells (HUVECs).

Elavarasan *et al.* prepared nano rod shaped triazine functional hierarchical mesoporous organic polymers (HMOP) containing Cu metal. This catalyst was used to synthesize triazole derivatives (20) *via* stirring at 80 °C in the presence of water as a solvent [46]. In the same year 2019, Gholampour *et al.* synthesized a library of 1,4-naphthoquinone-1,2,3-triazole hybrids (21) using $CuSO_4$ (0.15 mmol) and sodium ascorbate (.05 mmol) catalyzed click chemistry approach from 2-(prop-2-ynylamino)naphthalene-1,4-dione and different azidomethyl-benzene analogs. The anti-cancer activity of synthesized compounds was anticipated against three cancer cell lines (MCF-7, HT-29 and MOLT-4) by MTT assay. The compound **21f** possessed the highest activity [47]. In continuation to this, magnetic CuFe2O4/g-C3N4 hybrids were synthesized and their catalytic activity was examined in the synthesis

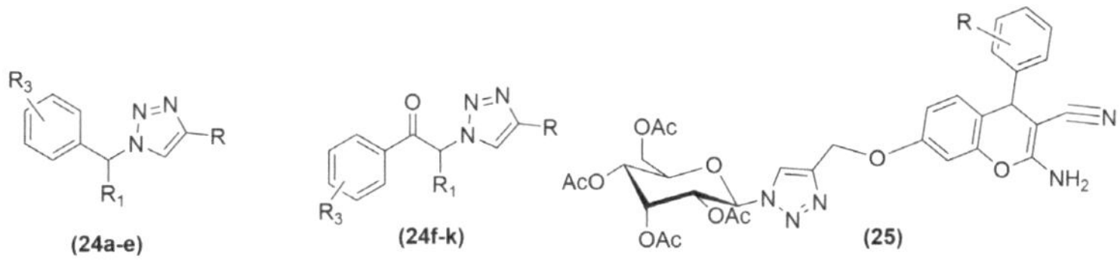

Figure 15.
Triazole derivatives.

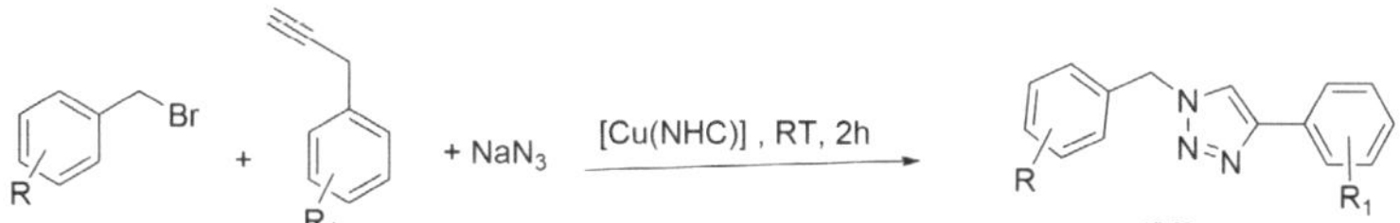

Figure 16.
Synthesis of triazole derivatives.

of triazole derivatives (22 and 23) using terminal alkyne, azide, epoxide or haloarene (**Figure 14**) [48].

Pourmohammad *et al*. synthesized (CuI@[PMMA-CO-MI]) (0.02 g) nano catalyst and employed them in the synthesis of triazole derivatives using terminal alkynes, α-haloketones or alkyl halides and sodium azide in H_2O at RT to give 1,4-disubstitued 1,2,3-triazoles (24) (**Figure 15**). The catalyst being heterogenous and regioselective gave high yields in short reaction times [49].

Thanh *et al*. [50] designed a hybrid structure of chromene and triazole by applying click chemistry approach for the synthesis of 1H-1,2,3-triazole-tethered 4H-chromene-D-glucose analogs (25) using Cu@MOF-5 (2 mol%) as a catalyst to afford 80–97.8% yields. The copper supported over metal organic frame work was found better catalyst in comparison to conventional catalysts *viz*. $CuSO_4.5H_2O$-sodium ascorbate, CuI, Cu Nps, $CuIM_2$(IM is imidazole) as it afforded high yields of desired products in less reaction time and in the presence of ethanol whereas other required long reaction time and non green solvent. All the derivatives were assessed for *in vitro* anti-microbial activity with MIC values in the range of 1.56–6.25 μM (**Figure 15**).

3.2.2 Synthesis of other organic molecules

Click chemistry has been used to synthesize biologically active hybrids of several synthetic organic molecules. In lieu of this, Sharova *et al.* [51] demonstrated click chemistry inspired phosphorylation of anabasine, camphor, and cytisine using Cu assisted 1,3-diploar cycloaddition reaction. Later on, Touj *et al.* [52] synthesized copper N-heterocyclic carbene (Cu-NHC) complexes using benzimidazolium salt as a catalyst. These complexes were further used for the synthesis of triazole derivatives (26) (**Figure 16**).

The reaction involved mild reaction conditions, water as a green solvent with low catalyst loading, no need of further purification which made the protocol eco-friendly. Bernard *et al.* [53] investigated a cost effective, and convenient click chemistry inspired synthesis of cyclooctyne (27) and trans-cyclooctene (28) using inexpensive Cu powder as a catalyst (**Figure 17**).

Qui *et al.* [54] synthesized parthenolide–thiazolidinedione (29) hybrids using click chemistry-mediated coupling. The compounds were screened for anti-proliferative activity against prostate (PC3), breast (MDA-MB-231), and human

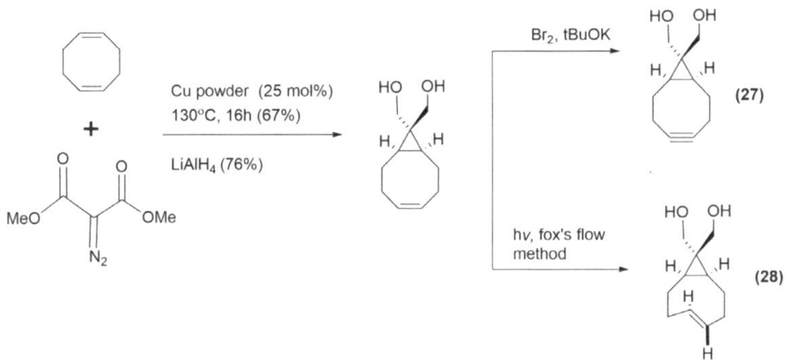

Figure 17.
Synthesis of cyclooctyne and trans-cyclooctene.

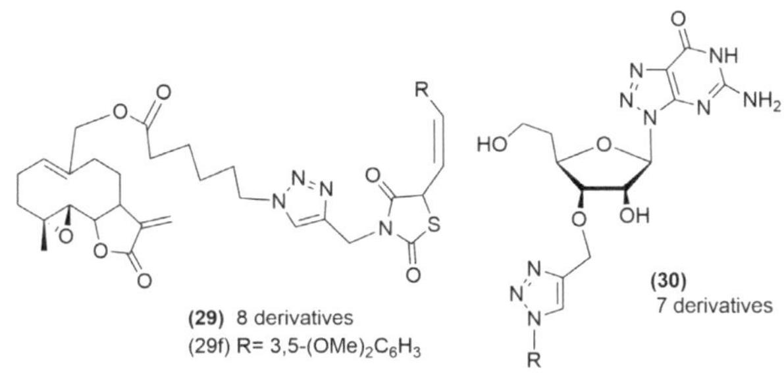

Figure 18.
Synthesis of parthenolide–thiazolidinedione and 3'-O-1,2,3-triazolyl-guanosine-5'-o-monophosphate derivatives.

erythroleukemia cell line (HEL) by MTT assay. The compound (29f) having 3,5-dimethoxyphenyl group exhibited the highest inhibitory effect against HEL (IC_{50} = 2.99 ± 0.22 μM), MDAMB-231 (IC_{50} = 2.07 ± 0.19 μM), and PC3 (IC_{50} = 3.09 ± 0.20 μM) (**Figure 18**).

Senthilvelan *et al.* [55] designed synthesis of 3'-O-1,2,3-triazolyl-guanosine-5'-*o*-monophosphate (30) from in situ generation of azide from the resultant bromide followed by copper and β-cyclodextrin catalyzed cyclo-addition with 3'-O-propargyl guanosine monophosphate in aqueous media. The designed pathway has high regioselectivity and gave good yield of products (**Figure 18**).

3.3 Metal-free approach for click reaction

Several new metal-free click chemistry assisted syntheses of heterocyclic scaffolds have been designed up to date. These pathways involve a variety of functional group tolerance in the substrate of cyclo-addition reaction. These synthetic pathways can be achieved under mild conditions and give high yields of desired products using organo-catalyst [56, 57]. In 2010, Fokin and co-authors developed the first transition metal-free synthesis of 1,5-diaryl-1,2,3-triazoles (31) employing azide-alkyne cyclo-addition [58] (**Figure 19 Method 1**). In this reaction, tetraalkylammonium hydroxide was used as the catalyst that provided moderate to high yield of products. Later on, in 2013, Ramachary and co-workers [59] achieved a region-selective synthesis of N-arylbenzotriazoles at room temperature using a cyclic enone and an arylazide under pyrrolidine catalysis at room temperature. Additional aromatization by DDQ gave fused heterocyclic scaffolds (32) (**Figure 19 Method 2**). In the subsequent year, a one-pot tandem,

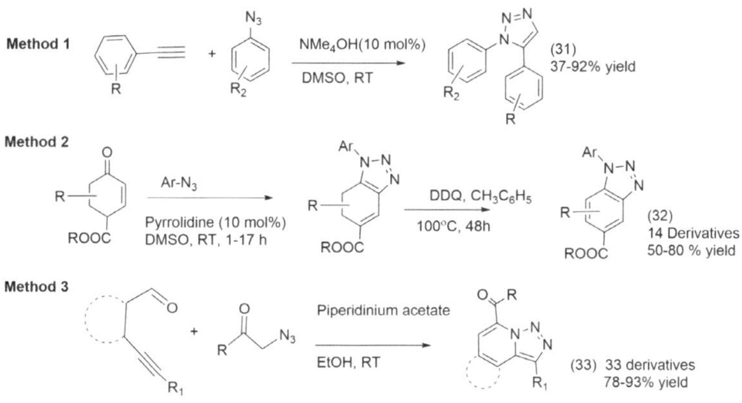

Figure 19.
Metal-free synthetic route of triazole based heterocycles.

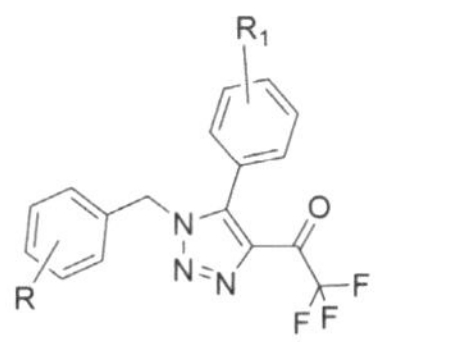

(34)
12 derivatives
(32b) R=4-CH_3, R_1=H

Figure 20.
4-Trifluroacetyl 1,2,3-triazole derivatives.

Knoevenagel/azide-alkyne cycloaddition reaction between indole, aromatic aldehydes or pyrazole and phenylazide was fruitfully accomplished in the presence of piperidinium acetate in methanol to furnish the desired triazole derivatives (33) (**Figure 19 Method 3**) [60].

In 2018, Han *et al.* [61] developed a metal-free and solvent-free approach for the synthesis of 4-trifluroacetyl 1,2,3-triazoles in good yields with great selectivity. The synthesized compounds were examined for the anti-cancer activity and compound **34b** possessed superior activity as compared to others against HePG2 (0.0267 μmol/mL) (**Figure 20**).

In the same year, Tan *et al.* [62] used thiol-ene click chemistry for the controlled functionalization of poly vinylidene fluoride in the presence of a base. The mechanism of the reaction suggests that it involves addition reaction followed by both Markonikov and anti-Markonikov mechanism and furnishes the same product. Later on in 2019, Moore and co-workers [63] designed a novel methodology for the synthesis of ionic liquids which were based on fluoroalkynyl imidazolium using thio-ene/yne click chemistry. The pathway has high conversion efficiency and high yields with no need for further purification.

3.4 Visible light assisted click chemistry

Visible-light-assisted organic transformations have received a huge response in chemical synthesis in order to design environmentally friendly approaches. The synthesis using economical, easily available visible-light sources have become vanguard in the synthetic chemistry as a prevailing approach for the activation of small molecules to furnish the desired products [64–66].

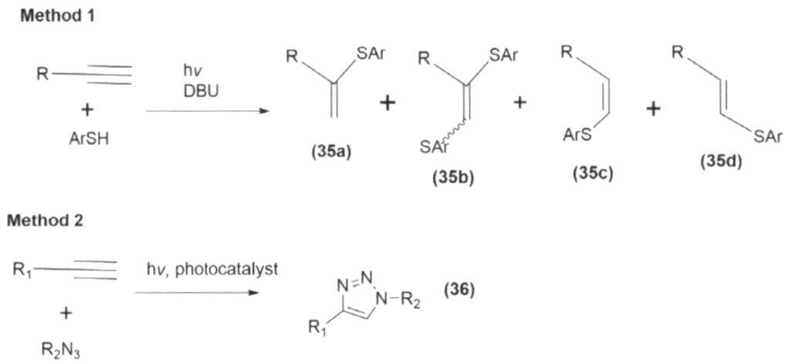

Figure 21.
Visible light assisted synthesis of vinyl sulphide.

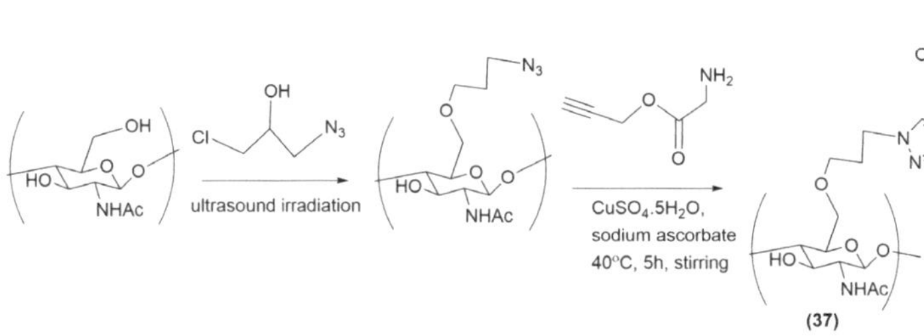

Figure 22.
Synthesis of ultrasound assisted 1-azido-3-chloropropan-2-ol azido chitin derivatives.

Burykina *et al.* [67] synthesized different kinds of vinyl sulphides **(35)** in high yields with good selectivity using thiol-yne click reaction using visible light. The designed pathway is transition-metal-free and gave Markonvikov-type product through a radical photo-redox pathway (**Figure 21 Method 1**).

Recently, Wu *et al.* [68] synthesized triazole analogs **(36)** through photo-redox electron-transfer mechanism. The authors inspected the reaction of benzyl azide with phenylacetylene using diverse photo-catalysts under ambient reaction conditions like room temperature (RT), air, and visible light irradiation. The catalyst $(piq)_2Ir(acac)$ or TPPT-Cl catalyzed the formation of triazole derivatives. The designed pathway is high region-selective, high yielding, having a high atom economy, and using solar catalysis **(Figure 21 Method 2).**

3.5 Ultrasound assisted click chemistry

Ultrasound assisted reactions are milder and faster. The mechanism of ultrasound is based on an acoustic cavitation phenomenon. This technology hastens the reaction in both heterogeneous and homogeneous media, due to amplified energy intake. It shortens the reaction time and augments the competence of the system by triggering the catalyst surface area and removing deposited impurities [69, 70].
A decades ago, Cintas *et al.* [71] depicted the synthesis of 1,4- disubstituted 1,2,3-triazole analogs using Cu under ultrasound irradiation exclusive of a ligand. Later on, a heterogeneous catalytic system, Cu(II) doped clay was used at RT with ultrasonic irradiations [72]. The use of heterogeneous catalyst evaded needless complexity due to copper (I) salt redox protocol that involved the presence of ligands and protecting agents. The reaction is eco-friendly, easy to prepare, and recoverable. One-pot synthesis of 1,4-disubstituted-1,2,3-triazoles was successfully achieved using a benzyl or alkyl halide, sodium azide, and a terminal alkyne under these conditions [73]. The formation of triazole starting from a TMS protected alkynylglycoside was also demonstrated under ultrasound conditions with *in situ*

deprotection of TMS group [74]. In 2020, Kritchenkov *et al*. synthesized triazole chitin derivatives and used them in the synthesis of Pd(II) complexes. Initially, ultrasound-assisted interaction of chitin with 1-azido-3-chloropropan-2-ol gave azido chitin and was further converted to triazole derivatives **(37)** that were used as ligands for the complex formation (**Figure 22**) [75].

3.6 Microwave-assisted click chemistry

The use of microwave irradiation in cyclo-addition reactions for click chemistry has also been comprehensively deliberated. It allows efficient internal heat transfer and therefore decreases the reaction time as well as enhances the reaction rate with high yield [76, 77]. The increased temperature can be used over short periods thus avoiding decomposition or polymerization. Ashok and co-workers demonstrated the synthesis of 1,2,3-triazole analogs using microwave irradiations in 8–10 min and examined their antimicrobial activity [78] (**Figure 23 Method 1**). This method has also been applied in the preparation of 1,2,3-triazole analogs of nucleosides [79] (**Figure 23 Method 2**). In general, those reactions which require prolonged conventional heating are accomplished in just 10–15 min using microwave irradiation. A chronological one pot Ru catalyzed cycloaddition was also designed from primary aryl or aliphatic bromides (**Figure 23 Method 3**) [80].

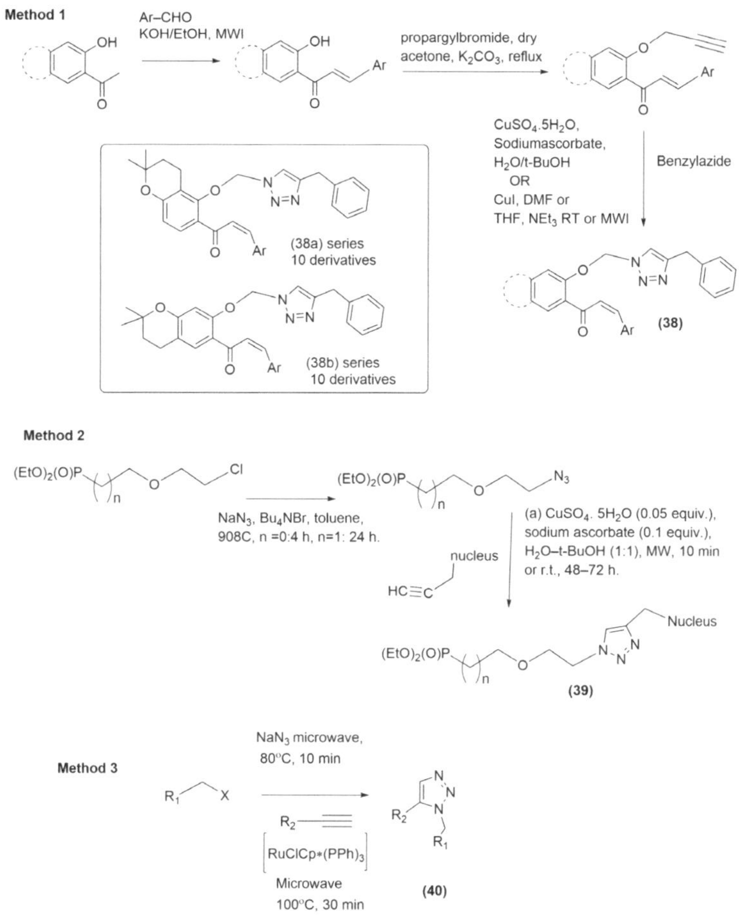

Figure 23.
Microwave assisted synthesis of triazole based scaffolds.

4. Click chemistry in polymer synthesis

In the past two decades, various polymers have been introduced through ionene synthesis, click chemistry, and Michael addition *via* polycondensation and polyaddition process. Click chemistry reactions are known as reliable, powerful, high-yielding, and selective for the synthesis of novel and combinatorial compounds *via* Diels Alder cyclo-additions, copper-catalyzed azide-alkyne cycloadditions (CuAAC), and azide nitrile cycloadditions process [81, 82]. In 2013, Pasini reviewed the utility of click reaction for the efficient synthesis of macrocyclic structures like polymers, bio-conjugates, and dendrimers in different contexts [83]. Recently, Arslan and Tasdelen systematically reviewed the applications of click chemistry in polymer design and synthesis, and studies based on their architecture like block, cyclic, star, hyperbranched, and graftbrush comb polymers [84].

In 2018, Acik and co-authors demonstrated a simple copper (I)-catalyzed azide-alkyne cyclo-addition "click" reaction for the synthesis of polypropylene-graft-poly (L-lactide) copolymers (PP-g-PLAs) using different feeding ratio of alkyne end-functionalized poly(L-lactide) azide and side-chain functionalized polypropylene in the presence of CuBr/PMDETA and CuAAC [85]. This polymer exhibited special characteristics like good thermal property, wettability and biodegradability.

Öztürk and companions introduced efficient click chemistry inspired synthesis of an amphiphilic copolymer **(41)** from the reaction of propargyl-PEG and terminally azidepoly(ε-caprolactone) in $CHCl_3$ at ambient temperature **(Figure 24)** [86]. This method displayed a synergistic arrangement of hydrophilic PEG and crystalline PCL to furnish novel materials with good applicability.

Yang *et al.* synthesized poly(3-hexylthiophene)-multiwalled carbon nanotube (P3HT-MWCNT) hybrid materials from in-situ click chemistry using Cu(I) / DBU catalytic system [87]. This novel hybrid also termed as organic–inorganic donor-acceptor material displayed special characteristics such as better thermal stability, higher melting point of 243.2 °C, good solubility, and optical properties.

Wang *et al.* reported a novel and efficient method for the synthesis of amphiphilic star-like rod-coil block copolymer poly(acrylic acid)-*block*poly(3-hexylthiophene) through the combined effect of atom transfer radical polymerization, quasi-living Grignard metathesis method, and thiol–ene click reaction to furnish narrow molecular weight distribution and well-defined molecular structures [88].

Agrihari *et al.* introduced CuAAC catalyzed synthesis of *p-tert*-butylcalix[4] arene linked benzotriazolyl dendrimers using $CuSO_4.5H_2O$ and NaN_3 to prepare N-1, N-2 type 6 fold compounds in good yields [89]. The synthesized compounds were evaluated for *in vitro* and *in vivo* anti-bacterial studies against a range of microbes and demonstrated good biological potential. Chen and his companions devised superhydrophobic cotton fabric from mercaptan and vinyl trimethoxysilane using ultraviolet irradiation *via* thiol-ene click chemistry [90]. This fabric possesses special characteristics like economic, highly resistant towards acids, acetone, UV light, water, and other liquids.

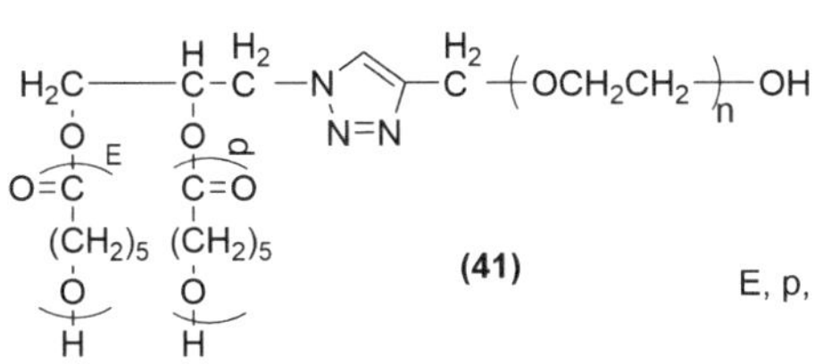

(41)

E, p, n are numbers

Figure 24.
Poly(CL-co-EG)star-type amphiphilic coploymer.

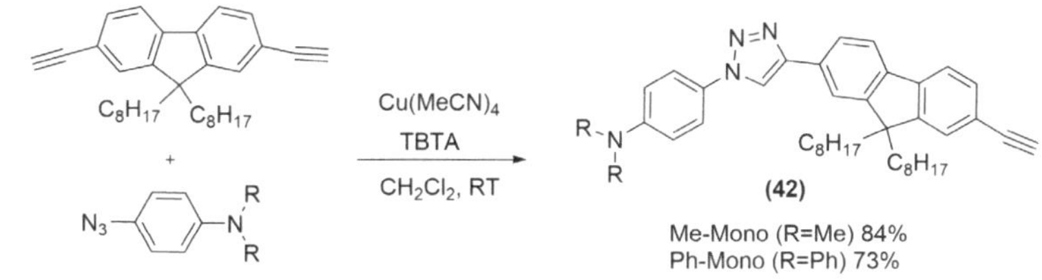

Figure 25.
Synthesis of triazole-based photo-initiators.

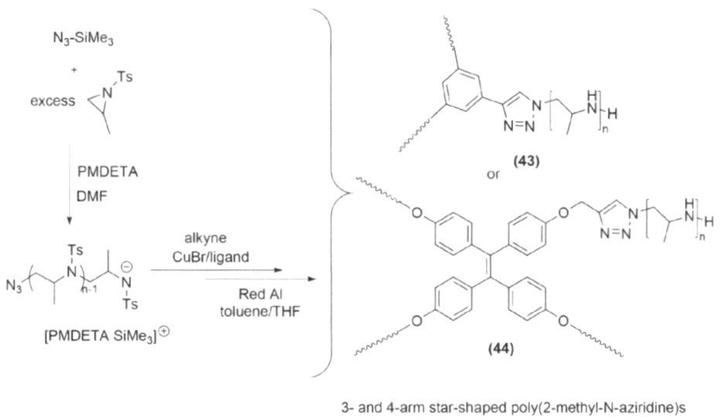

Figure 26.
3-and 4-arm star-shaped poly(2-methyl-N-aziridine)s.

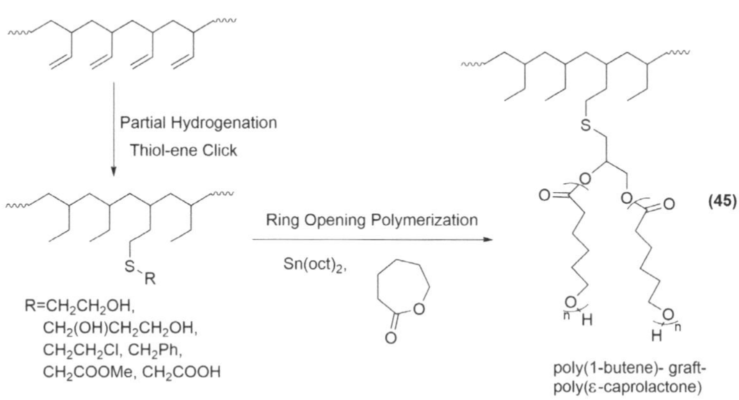

Figure 27.
Functionalized poly(1-butene) synthesis.

Henning *et al*. utilized copper-catalyzed azide/alkyne cycloaddition reaction for the efficient synthesis of triazole-based photo-initiators **(42)** for the two-photon polymerization process **(Figure 25)** [91]. Here Me-Mono and Ph-Mono initiators displayed higher tolerability and sensitivity in microfabrication areas.

A novel, facile and efficient synthesis of 3- and 4-arm star-shaped poly (2-methyl-N-aziridine)s **(43, 44)** from ring opening reaction of N-sulfonyl aziridines in the presence of trimethylsilylazide and PMDETA (N,N,N′,N″,N″-pentamethyldiethylenetriamine) through click reaction with CuBr and alkyne was demonstrated by Luo *et al*. **(Figure 26)** [92].

Cai *et al.* presented a one-step click chemistry process for the synthesis of high performance graphene oxide/ styrene-butadiene rubber (GO/SBR) composites using pentaerythritoltetra(3-mercapto propionate) [93]. Experiments and molecular simulation results concluded that these composites displayed upgraded gas permeability, thermal conductivity, dynamic, and static mechanical performances.

Tian and companions demonstrated the synthesis of the functionalized poly(1-butene)**(45)** *via* sequential thiol-ene click reaction and ring-opening polymerization using poly(1,3-butadiene) as a substrate **(Figure 27)** [94]. Here C=C bond was further functionalized from thiol-ene reaction using hydroxyl-containing thiol compounds.

Zhang *et al*. devised synthesis of thiol-maleimide 'click' chemistry based β-cyclodextrin polymers in an aqueous medium without generating by-products [95]. The structure of products was affected by temperature range *i.e.* higher temperature gave higher molecular weighted and compact structures. The obtained polymers showed better dissolution performance and drug complex-forming capacity as compared to the parental structure.

Gao *et al*. reported thiol-ene click reactions in polysulfide oligomers and acrylate monomers to prepare processable and self-healable thermosets and elastomers **(46)** *via* different pathways like photo-initiator, redox-initiator system and base mediated catalytic approaches **(Figure 28)** [96]. Reprocessable and self-healable properties depend upon polymer structure and their synthetic methodology, therefore, DBU based catalytic synthesis displayed better activity as compared to other processes, due to their catalytic efficiency for disulfide bond exchanges.

Zhu *et al*. introduced facile click chemistry assisted poly(2,2,6,6-tetramethylpiperidin-1-oxyl-4-yl methacrylate) graphene oxide composite (PTMA-GO) **(47)** assisted reaction in ambient conditions and utilized them as cathode materials **(Figure 29)** [97]. After completing 300 charge–discharge cycles,the specific capacity was found to be 2.3 times higher for this composite as compared to the PTMA electrode.

Shen *et al.* devised synthesis of superhydrophilic and superoleophobic PEGylated PAN membrane **(48)** from poly (ethylene glycol) methyl ether methacrylate (PEGMA) monomers *via* thiol-ene click chemistry **(Figure 30)** [98]. After this fabrication, pore size of the membrane was reduced and displayed low flux. The

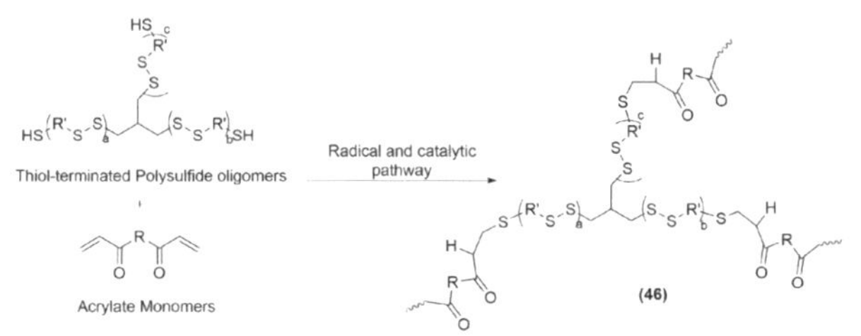

Figure 28.
Process able and self-healable polymer synthesis.

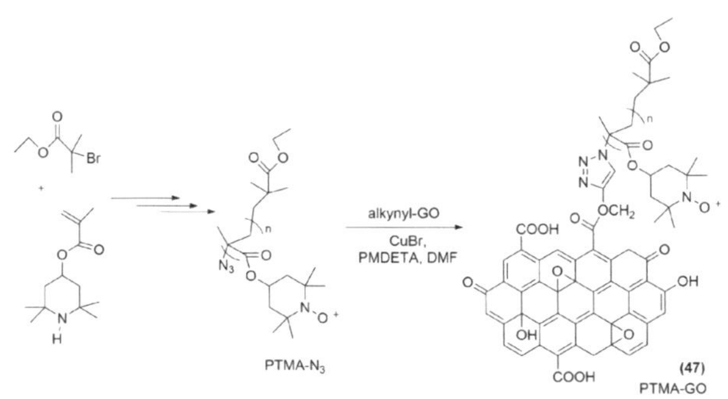

Figure 29.
PTMA-GO polymer synthesis.

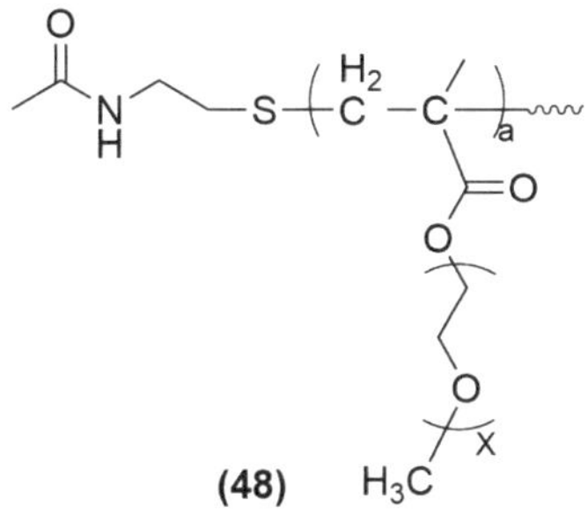

(48)

PEGylated PAN membranes

Figure 30.
PEGylated PAN membranes.

fabricated membrane exhibited excellent separation (99%) of various oil–water emulsion and fouling-resistant ability.

5. Conclusion

Synthetic organic chemistry includes the synthesis of biologically active molecules and designing of potent scaffolds. Click chemistry is one of the toolboxes for chemistry, biology, nano, and material sciences. It has vivid applications in the synthesis of organic molecules, polymers, nanoparticles, biosensors, and many more. The concept of click chemistry fulfills the green aspects of a reaction. In this chapter, we have deliberated an incredible flurry of activities in the field of click chemistry inspired synthesis. This study highlights the current advancements in the synthesis of heterocyclic and other cyclic structures using click reactions. The insertion of a triazole ring with the help of click reaction increases the biological activity of the synthesized compounds. Different pathways with metal or metal-free conditions using conventional or non-conventional reaction methods have also been demonstrated in this chapter.

Acknowledgements

The authors are grateful to the Department of Chemistry, Mohan Lal Sukhadia University, Udaipur (Raj.), India for providing necessary library facilities for carrying out the work. A. Sethiya is thankful to UGC-MANF (201819MANF-2018-2019-RAJ-91971) for providing Senior Research Fellowship to carry out this work. N. Sahiba is very much grateful to CSIR-Delhi (file no. 09/172(0088) 2018-EMR-I), New Delhi for providing Senior Research Fellowship as a financial support.

Conflict of interest

The authors declare no conflict of interest.

Author details

Ayushi Sethiya, Nusrat Sahiba and Shikha Agarwal*
Synthetic Organic Chemistry Laboratory, Department of Chemistry, MLSU, Udaipur, India

*Address all correspondence to: shikhaagarwal@mlsu.ac.in

References

[1] Balasubramanian G, Balasubramanian KK. Applications of click chemistry in drug discovery and development. In: Click reactions in organic synthesis. Wiley;2016. DOI.10.1002/9783527694174

[2] Hein D, Liu X-M, Wang D. Click chemistry, a powerful tool for pharmaceutical sciences. Christopher Pharm Res. 2008;25(10):2216–2230. DOI:10.1007/s11095-008-9616-1.

[3] Hou J, Liu X, Shen J, Zhao G, Wang PG. The impact of click chemistry in medicinal chemistry. Expert Opin Drug Discov. 2012;489–501. DOI.10.1517/17460441.2012.682725.

[4] Lauwaet T, Miyamoto Y, Le SIC, Kalisiak J, Korthals KA, Ghassemian M, Smith DK, Sharpless KB, Fokin VV, Eckmann L. Click chemistry-facilitated comprehensive identification of proteins adducted by antimicrobial 5-nitroimidazoles for discovery of alternative drug targets against giardiasis. PLOS Negl Trop Dis. 14(4): e0008224. https://doi.org/10.1371/journal.pntd.0008224

[5] Li P-Z, Wang X-J, Zhao Y. Click chemistry as a versatile reaction for construction and modification of metal-organic frameworks. Coord Chem Rev. 2019;380:484–518.

[6] Tarnowska M. Krawczyk T. Click chemistry as a tool in biosensing systems for sensitive copper detection. Biosens Bioelectron. 2020;169:112614

[7] Li S, Zhou J, Huang YH, Roy J, Zhou N, Yum K, Sun X, Tang L. Injectable click chemistry-based bioadhesives for accelerated wound closure. ActaBiomater. 2020;110:95–104.

[8] Ouyang T, Liu X, Ouyang H, Ren L. Recent trends in click chemistry as a promising technology for virus-related research. Virus Res. 2018;256:21–28.

[9] Huang C-J. Advanced surface modification technologies for biosensors. In: Chemical, gas, and biosensors for the internet of things and related applications. Elsevier;2019. p. 65–86. DOI.org/10.1016/B978-0-12-815409-0.00005-X

[10] Fang-Ling Z, Si-Yu G, Yuan-Dong X, Jin-Ming Z, Yi L, Ning L. Applications of click chemistry reaction for proteomics analysis. Chinese J Anal Chem. 2020;48(4):431–438.

[11] Meyer JP, Adumeau P, Lewis JS, Zeglis BM. Click chemistry and radiochemistry: The first 10 years. Bioconjug Chem. 2016;27(12):2791–2807.

[12] Kenry. Bio-orthogonal click chemistry for in vivo bioimaging. Trends Chem. 2019;1(8):763–778.

[13] Yang H, Sukamtoh E, Du H, Wang W, Ando M, Kwakwaa YN, Zhang J, Zhang G. Click chemistry approach to characterize curcumin-protein interactions in vitro and in vivo. J Nutr Biochem. 2019;68:1-6. DOI.org/10.1016/j.jnutbio.2019.02.010.

[14] Binder WH, Sachsenhofer R. Click' chemistry in polymer and materials science. Macromole Rapid Commun. 2007;28(1):15–54. https://doi.org/10.1002/marc.200600625.

[15] Moses JE, Moorhouse AD. The growing applications of click chemistry, ChemSoc Rev. 2007;36(8):1249–1262.

[16] Hanson J. John Cornforth(1917–2013). Nature. 2014;506:35.

[17] Shirame SP, Bhosale RB. Green approach in click chemistry. InTech

Open. 2018. http://dx.doi.org/10.5772/intechopen.72928.

[18] Kolb HC, Finn MG, Sharpless KB. Click chemistry: Diverse chemical function from a few good reactions. Angew Chem Int Ed. 2001;40:2004–2021.

[19] Chandrasekaran S, Ramapanicker R. Click chemistry route to the synthesis of unusual amino acids, peptides, triazole-fused heterocycles and pseudodisaccharides. Chem Rec. 2017;17 (1):63–70. DOI: 10.1002/tcr.201600093.

[20] Michinobu T, ÅoisDiederich F. The [2+2] cycloaddition-retroelectrocyclization (ca-re) click reaction: Facile access to molecular and polymeric push-pull chromophores. Angew Chem Int Ed. 2018;57:2–28.

[21] Totobenazara J, Burke AJ. New click-chemistry methods for 1,2,3-triazoles synthesis: Recent advances and applications. Tetrahedron Lett. 2015;56 (22):2853–2859. Doi: http://dx.doi.org/10.1016/j.tetlet.2015.03.136

[22] Meldal M, Diness F. Recent fascinating aspects of the CuAAC click reaction. Trends Chem. 2020,2(6):569–584. https://doi.org/10.1016/j.trechm.2020.03.007.

[23] Heravi MM, Zadsirjan V, Dehghani M, Ahmadi T. Towards click chemistry: Multicomponent reactions via combinations of name reactions. Tetrahedron 2018;74:3391–3457.

[24] Huisgen R. 1,3-Dipolar cycloadditions. Past and future. Angew Chem Int Ed Engl. 1963;2:565–598.

[25] Wang C, Ikhlef D, Kahlal S, Saillard J-Y, Astruc D. Metal-catalyzed azide-alkyne "click" reactions: Mechanistic overview and recent trends. Coord. Chem. Rev. 2016;316:1–20.

[26] Johansson JR, Beke-Somfai T, Stalsmeden AS, Kann. Ruthenium-catalyzed azide alkyne cycloaddition reaction: scope, mechanism, and applications. Chem. Rev. 2016;116(23): 14726–14768.

[27] Meng X, Xu XY, Gao TT, Chen B. Zn/C-catalyzed cycloaddition of azides and aryl alkynes. Eur. J. Org. Chem. 2010;5409–5414.

[28] McNulty J, Keskar K, Vemula R. The first well-defined silver(i)-complex-catalyzed cycloaddition of azides onto terminal alkynes at room temperature. Chem. Eur. J. 2011;17:14727–14730.

[29] Kim UB, Jung DJ, Jeon HJ, Rathwell K, Lee SG. Synergistic dual transition metal catalysis. Chem. Rev. 2020;120(24):13382–13433.

[30] Tăbăcaru A, Furdui B, Ghinea IO, Cârâc G, Dinică RM. Advances in click chemistry reactions mediated by transition metal based systems. InorganicaChimicaActa. 2017;455(2): 329–349.

[31] Rostovtsev VV, Green LG, Fokin VV, Sharpless KB. A stepwise huisgen cycloaddition process: Copper (I)-catalyzed regioselective "ligation" of azides and terminal alkynes. Angew Chem Int Ed. 2002;41:2596–2599.

[32] Guo H, Yanga F, Zhanga Y, Di XD. Facile synthesis of mono- and polytopicβ-cyclodextrin aromatic aldehydes by click chemistry. Synth Commun. 2014;45(3):338–347. doi.org/10.1080/00397911.2014.963400.

[33] Kumar S. Design and synthesis of 2′-d.eoxy-2′-[(1,2,3)triazol-1-yl]uridines using click chemistry approach. Nucleosides, Nucleotides, Nucleic Acids, 2015,34(5):371–378. DOI: 10.1080/15257770.2014.1003652.

[34] Tale RH, Gopula VB, Toradmal GK. "Click" ligand for "Click" chemistry: (1-(4-Methoxybenzyl)-1-H-1, 2, 3-triazol-4-yl) methanol (MBHTM) accelerated

copper-catalyzed [3+2] azide-alkyne cycloaddition (CuAAC) at low catalyst loading. Tetrahedron Lett. 2015; http://dx.doi.org/10.1016/j.tetlet.2015.09.010.

[35] Shamala D, Shivashankara K, Chandra, Mahendra M. Synthesis of N1 and N2 coumarin substituted 1,2,3-triazole isomers via click chemistry approach. Synth Commun. 2016;46(5): 433–441. doi.org/10.1080/00397911.2016.1140785

[36] Yarlagadda B, Devunuri N, Mandava VBR. Facile synthesis of n-(benzyl-1h-1,2,3-triazol-5-yl) methyl)-4-(6-methoxybenzo [d] thiazol-2-yl)-2-nitrobenzamides via click chemistry. J Heterocycl Chem. 2017;54(2):864–870.

[37] Anand A, Kulkarni MV. Click chemistry approach for the regioselective synthesis of iso-indoline-1,3-dione-linked 1,4 and 1,5 coumarinyl 1,2,3-triazoles and their photophysical properties. Synth Commun. 2017;47(7): 722–733.

[38] Anandhan R, Kannan A, Rajakumar P. Synthesis and anti-inflammatory activity of triazole-based macrocyclic amides through click chemistry. Synth Commun. 2017;47(7): 671–679. DOI: 10.1080/00397911.2016.1254800.

[39] Lia J, Chen D, Zhang D, Wang Y, Yu Y, Gao L, Huang M. Preparation of triazole compounds via click chemistry reaction and formation of the protective self-assembled membrane against copper corrosion. Colloids and Surfaces A. 2018; 550:145–154.

[40] Kumar H, Savanur M, Naik KN, Ganapathi SM, Kim KM, Kalkhambkar RG. Click chemistry inspired design, synthesis and molecular docking studies of coumarin, quinolinone linked 1,2,3-triazoles as promising anti-microbial agents. ChemistrySelect 2018;3:5296–5303.

[41] Yarovaya O, Artyushin OI, Moiseeva AA, Zarubaev VV, Slita AV, Galochkina AV, Muryleva AA, Borisevich SS, Salakhutdinov NF, Brel VK. Synthesis of camphecene and cytisine conjugates using click chemistry methodology and study of their antiviral activity. Chem Biodiver. 2019;16(11):e1900340. DOI.10.1002/cbdv.201900340

[42] Khanapurmath N, Kulkarnia MV, Joshi S.D., Kumar G.N.A. A click chemistry approach for the synthesis of cyclic ureido tethered coumarinyl and 1-aza coumarinyl 1,2,3-triazoles as inhibitors of *Mycobacterium tuberculosis* H37Rv and their in silico studies. Bioorg Med Chem. 2019;27:115054.

[43] Chetia M, Gehlot PS, Kumar A, Sarma D. A recyclable/reusable hydrotalcite supported copper nano catalyst for 1,4-disubstituted-1,2,3-triazole synthesis via click chemistry approach. Tetrahedron Lett. 2018;59: 397–401.

[44] Poshala S, Thung S, Manchala S, Kokatla HP. In situ generation of copper nanoparticles by rongalite and their use as catalyst for click chemistry in water. ChemistrySelect 2018;3:13759–13764.

[45] Chavana PV, Desai UV, Wadgaonkar PP, Tapase SR, Kodam KM, Choudhari A, Sarkar D. Click chemistry based multicomponent approach in the synthesis of spirochromenocarbazole tethered 1,2,3-triazoles as potential anticancer agents. Bioorg Chem. 2019;85:475–486.

[46] Elavarasan S, Bhaumik A, Sasidharan M. An efficient Cu-mesoporous organic nanorod for Frieldländer quinoline synthesis, and click reactions. Chem Cat Chem. 2019; 11(17):4350–4350. DOI:10.1002/cctc.201900860.

[47] Gholampour M, Ranjbar S, Edrakic N, Mohabbatic M, Firuzi O,

Khoshneviszadeh M. Click chemistry-assisted synthesis of novel aminonaphthoquinone-1,2,3-triazole hybrids and investigation of their cytotoxicity and cancer cell cycle alterations. Bioorg Chem. 2019;88: 102967.

[48] Khalili D, Rezaee M. Impregnated copper ferrite on mesoporous graphitic carbon nitride: An efficient and reusable catalyst for promoting ligand-free click synthesis of diverse 1,2,3-triazoles and tetrazoles. Appl Organo Metal Chem. 2019;e5219. https://doi.org/10.1002/aoc.5219

[49] Pourmohammad N, Heravi MM, Ahmadi S, Hosseinnejad T. In situ preparation and characterization of novel CuI-functionalized poly[(methyl methacrylate)-co-maleimide] as an efficient heterogeneous catalyst in the regioselective synthesis of 1,2,3-triazoles via click reaction: Experimental and computational chemistry. Appl Organo Metal Chem. 2019;33(7):e4967.

[50] Than ND, Hai DS, Bich VTN, Hien PTTH, Duyen NTD, Mai NT, Dung TT et al. Efficient click chemistry towards novel 1H-1,2,3-triazole-tethered 4Hchromene D-glucose conjugates: Design, synthesis and evaluation of in vitro antibacterial, MRSA and antifungal activities. Eur J Med Chem. 2019; 167: 454–471.

[51] Sharova EV, Genkina GK, Vinogradova NM, Artyushin OI, Yarovaya OI, Brel VK, Phosphorylation of natural products—Cytisine, anabasine, and camphor using click chemistry methodology. Phosphorus Sulfur and Silicon Relat Elem 2016;191 (11–12):1556–1557. DOI: 10.1080/10426507.2016.1213257

[52] Touj N, Özdemir I, Yaşar S, Hamdi N. An efficient (NHC) Copper (I)-catalyst for azide–alkyne cycloaddition reactions for the synthesis of 1,2,3-trisubstituted triazoles: click chemistry. Inorg Chim Acta 2017;467: 21–32.doi: http://dx.doi.org/10.1016/j.ica.2017.06.065

[53] Bernard S, Kumar RA, Porte K, Thuery P, Taran F, Audisio D. A practical synthesis of valuable strained eight-membered-ring derivatives for click chemistry. Eur J Org Chem. 2018; 2000–2008. 10.1002/ejoc.201800139.

[54] Qiu J, Yuan C-M, Wen M, Li Y-N, Chen J, Jian J-Y, Huang L-J, Gu W, Li YM, Hao X-J. Design, synthesis, and cytotoxic activities of novel hybrids of parthenolide and thiazolidinedione via click chemistry. J Asian Nat Prod Res. 2020;22(5):425–433. DOI: 10.1080/10286020.2019.1597055

[55] Senthilvelan A, Shanmugasundaram M, Kore AR. An efficient synthesis of 3′-O-triazole modified guanosine-5′-O-monophosphate using click chemistry. Nucleosides, Nucleotides, Nucleic Acids 2019;38(6):418–427. DOI: 10.1080/15257770.2018.1554223.

[56] Becer CR, Hoogenboom R, Schubert US. click chemistry beyond metal-catalyzed cycloaddition. Angew. Chem. Int. Ed. 2009;48:4900–4908. DOI: 10.1002/anie.200900755.

[57] Dervaux B, Du Prez FE. Heterogeneousazide–alkyne click chemistry: Towards metal-free end Products. Chem. Sci. 2012;3:959–966. DOI: 10.1039/c2sc00848c.

[58] Kwok WS, Fotsing RJ, Fraser JR, Rodionov OV, Fokin VV. Transition-metal-free catalytic synthesis of 1,5-diaryl-1,2,3-triazoles. Org Lett. 2010; 12: 4217–4219.

[59] Ramachary DB, Shashank BA. Organocatalytictriazole formation, followed by oxidative aromatization: Regioselective metal-free synthesis of benzotriazoles. ChemEur J. 2013;19: 13175–13181.

[60] Maurya RA, Adiyala PR, Chandrasekhar D, Reddy CN, Kapure JS, Kamal A. Rapid access to novel 1,2,3-triazolo-heterocyclic scaffolds via tandem knoevenagel condensation/azide–alkyne 1,3-dipolar cycloaddition reaction in one pot. ACS Comb Sci. 2014;16:466–477.

[61] Han J, Ran J-X, Chen X-P, Wang Z-H, Wu F-H. Study on the green click-chemistry synthesis of 4-trifluoroacetyl-1,2,3-triazoles. Tetrahedron 2018; 74: 6985–6992.

[62] Tan S, Li D, Zhang Y, Niu Z, Zhang Z. Base catalyzed thiol–ene click chemistry toward inner -CH=CF-bonds for controlled functionalization of Poly (vinylidene fluoride). Macromol Chem Phys. 2018;1700632, DOI: 10.1002/macp.201700632.

[63] Moore LMJ, Greeson KT, Redeker ND et al. Fluoroalkylfunctional imidazoles and imidazolium–based ionic liquids prepared via thiol-ene/yne click chemistry. J Mol Liq. 2019;295:111677. https://doi.org/10.1016/j.molliq.2019.111677.

[64] Liu Q, Wu L-Z. Recent advances in visible-light-driven organic reactions. Natl Sci Rev. 2017;4(3):359–380.https://doi.org/10.1093/nsr/nwx039

[65] Yu X-Y, Chen J-R, Xiao W-J. Visible light-driven radical-mediated c–c bond cleavage/functionalization in organic synthesis. Chem. Rev. 2020, https://doi.org/10.1021/acs.chemrev.0c00030

[66] Jagan MR, Narayanama, Stephenson CRJ. Visible light photoredox catalysis: applications in organic synthesis. **Chem. Soc. Rev.** 2011;40:102–113.

[67] Burykina JV, Shlapakov NS, Gordeev EG, BurkhardK¨onig, Ananikov VP. Selectivity control in thiol–yne click reactions via visible light induced associative electron upconversion. Chem Sci. 2020;11: 10061–10070.

[68] Wu Z-G, Liao X-J, Yuan L, Wang Y, Zheng Y-X, Zuo J-L, Pan Y. Visible-light-mediated click chemistry for highly regioselective azide-alkyne cycloaddition via photoredox electron-transfer strategy. Chem Eur J. 2020;26 (25):5694–5700. 10.1002/chem.202000252.

[69] BaigRBN, Varma RS. Alternative energy input: Mechanochemical, microwave and ultrasound-assisted organic synthesis. Chem. Soc. Rev. 2012; 41:1559–1584.

[70] Banerjee B. Recent developments on ultrasound-assisted one-pot multicomponent synthesis of biologically relevant heterocycles. Ultrason Sonochem 2017;35:15–35.

[71] Cravotto G, Fokin VV, Garella D, Binello A, Boffa L, Barge A. Ultrasound-promoted copper-catalyzed azide—alkyne cycloaddition. J Comb Chem. 2010;12(1):13–15.

[72] Cintas P, Barge A, Tagliapietra S, Boffa L, Cravotto G.Alkyne–azide click reaction catalyzed by metallic copper under ultrasound. Nature Protocols 2010;5:607–616.

[73] Dar BA, Bhowmik A, Sharma A, Sharma PR, Lazar A, Singh AP, Sharma, M, Singh B. Ultrasound promoted efficient and green protocol for the expeditious synthesis of 1, 4 disubstituted 1, 2, 3-triazoles using Cu (II) doped clay as catalyst. Appl Clay Sci. 2013; 80–81: 351–357.

[74] Stefani HA, Silva NCS, Manarin F, Lüdtke DS, Zukerman-Schpector J. Madureira LS, Tiekink ERT. Synthesis of 1,2,3-triazolylpyranosides through click chemistry reaction. Tetrahedron Lett. 2012;53(14):1742–1747.

[75] Kritchenkova AS, Kletskov AV, Egorov AR, Kurasova MN, Tskhovrebov AG, Khrustalev VN. Ultrasound and click chemistry lead to a new chitin chelator. Its Pd(II) complex is a recyclable catalyst for the Sonogashira reaction in water. Carbohydr Polym. 2021;252:117167.

[76] Gupta AK, Singh N, Singh KN. Microwave assisted organic synthesis: Cross coupling and multicomponent reactions. Curr Org Chem. 2013;17(5). DOI : 10.2174/1385272811317050005

[77] Sharma N, Sharma UK, Van der Eycken EV. Microwave-assisted organic synthesis: overview of recent applications. In: Zhang W, Cue BW. Green techniques for organic synthesis and medicinal chemistry, 2nd Ed. Wiley; 2018.https://doi.org/10.1002/9781119288152.ch17,

[78] Ashok D, Gandhi DM, Srinivas G, Kumar V. Microwave-assisted synthesis of novel 1,2,3-triazole derivatives and their antimicrobial activity. Med Chem Res. 2014;23:3005–3018.

[79] Głowacka IE, Balzarini J, Wróblewski AE. Synthesis and biological evaluation of novel 1,2,3-triazolonucleotides. Arch Pharm Chem Life Sci. 2013;346:278–291.

[80] Johansson JR, Lincoln P, Nordén B, Kann N. Sequential one-pot ruthenium-catalyzed azide−alkyne cycloaddition from primary alkyl halides and sodium azide. J Org Chem. 2011;76(7):2355–2359.

[81] Zhang M, June SM, Long TE, Kong J. Principles of step-growth polymerization (polycondensation and polyaddition). In: Polymer Science: A comprehensive reference, 10 Volume Set. Vol. 5. Elsevier; 2011. p. 7–47. DOI: 10.1016/B978-0-12-803581-8.01410-7

[82] Milanese C. Chapter 5. Cycloaddition reactions in material science. In: Quadrelli P, editors. Modern applications of cycloaddition chemistry. Elsevier; 2019. p. 269. DOI: https://doi.org/10.1016/B978-0-12-815273-7.00005-8

[83] Pasini D. The click reaction as an efficient tool for the construction of macrocyclic structures. Molecules 2013; 18:9512–9530. DOI:10.3390/molecules18089512

[84] Arslan M, Tasdelen MA. Click chemistry in macromolecular design: Complex architectures from functional polymers. Chemistry Africa. 2019; 2: 195–214. DOI: https://doi.org/10.1007/s42250-018-0030-8

[85] AcikG,AltinkokC,Tasdelen MA. Synthesis and characterization of polypropylene-graft-poly(L-Lactide) copolymers by CuAAC click chemistry. J Polym Sci Part A: Polym Chem. 2018; 56(22):2595–2601.

[86] Öztürk T, Kılıçlıoğlu A, Savaş B, Hazer B. Synthesis and characterization of poly(ε-caprolactone-co-ethylene glycol) star-type amphiphilic copolymers by "click" chemistry and ring-opening polymerization. J MacromolSci A. 2018;55(8):588–594. DOI: 10.1080/10601325.2018.1481344

[87] Yang Z, Yu J, Fu K, Tang F. Preparation and characterization of poly (3- hexylthiophene) / carbon nanotubes hybrid material via in-situ click chemistry. Mater Chem Phys. 2018;223: 797–804. DOI: https://doi.org/10.1016/j.matchemphys.2018.11.050.

[88] Wang X, Zhao C, Li Y, Lin Z, Xu H. A Facile and highly efficient route to amphiphilic star-like rod-coil block copolymer via a combination of atom transfer radical polymerization with thiol–ene click chemistry. Macromol Rapid Commun. 2020;1900540.

[89] Agrihari AK, Singh M, Singh AS, Singh AK, Yadav S, Maji P, Rajkhowa S,

Prakash P, TiwariVK.Click inspired synthesis of p-tert-butyl calix[4] arenetetheredbenzotriazolyl dendrimers and their evaluation asanti-bacterial and anti-biofilm agents. New J Chem. 2020; 44:19300–19313. DOI: 10.1039/D0NJ02591G

[90] Chen X, Chu R, Xing T, Chen G. One-step preparation of superhydrophobic cotton fabric based on thiol-ene click chemistry. Colloid Surface A. 2020;125803. DOI: https://doi.org/10.1016/j.colsurfa.2020.125803

[91] Henning I, Woodward AW, Rance JA, Paul BT, Wildman RD, Irvine DJ, Moore JC. A click chemistry strategy for the synthesis of efficient photoinitiators for two-photon polymerization. Adv Funct Mater. 2020; 2006108.

[92] Luo W, Wang Y, Jin Y, Zhang Z, Wu C. One-pot tandem ring-opening polymerization of N-sulfonyl aziridines and "click" chemistry to produce well-defined star-shaped polyaziridines. J Polym Sci. 2020;1–10. DOI: 10.1002/pol.20200154

[93] Cai F, You G, Luo K, Zhang H, Zhao X, Wu S. Click chemistry modified graphene oxide/styrene-butadiene rubber composites and molecular simulation study. Compos Sci Technol. 2020;190:108061.

[94] Tian L, Gu J, Zhang H, Dong B. Preparation of functionalized poly(1-butene) from 1,2-polybutadiene via sequential thiol-ene click reaction and ring-opening polymerization. RSC Adv. 2020;10:42799–42803.

[95] Zhang M, Wei X, Xu X, Jin Z, Wang J. Synthesis and characterization of water-soluble β-cyclodextrin polymers via thiol-maleimide 'click' chemistry. Eur Polym J. 2020;128: 109603. DOI:10.1016/j.eurpolymj.2020.109603

[96] Gao H, Sun Y, Wang M, Wu B, Han G, Jin L, Zhang K, Xia Y. Self-healable and reprocessable acrylate-based elastomers with exchangeable disulfide crosslinks by thiol-ene click chemistry. Polymer, 2020;123132. https://doi.org/10.1016/j.polymer.2020.123132.

[97] Zhu J, Zhu T, Tuo H, Yan M, Zhang W, Zhang G, Yang X. TEMPO-contained polymer grafted onto graphene oxide via click chemistry as cathode materials for organic battery. Macromol Chemb Phys. 2020;2000160. DOI: 10.1002/macp.202000160

[98] Shen X, Liu P, He C, Xia S, Liu J, Cheng F, Suo H, Zhao Y, Chen L. Surface PEGylation of polyacrylonitrile membrane via thiol-ene click chemistry for efficient separation of oil-in-water emulsions. Sep Purif Technol. 2021;255: 117418.

Chapter 6

Anion-π Catalysis: A Novel Supramolecular Approach for Chemical and Biological Transformations

Ishfaq Ahmad Rather and Rashid Ali

Abstract

Catalysts by virtue of lowering the activation barrier helps in the completion of a chemical reaction in a lesser amount of time without being themselves consumed. Utilizing the diverse non-covalent interactions in the design and construction of catalysts, recently anion-π interactions were also introduced, giving rise to an emerging field of anion-π catalysis. In the newly constructed anion-π catalysts, significant lowering of activation energy occurs by virtue of anion-π interactions. Till now, several important reactions generating chiral centers have been carried out on the π-acidic surfaces of anion-π catalysts, thereby revealing the significance of anion-π catalysis in the domain of asymmetric synthesis. The motive of this chapter is to highlight the role of anion-π catalysis in asymmetric synthesis and we surely believe that it will offer new opportunities in supramolecular chemistry.

Keywords: anion-π catalysis, anion-π interaction, asymmetric synthesis, cascade reaction, enolate chemistry, naphthalene diimide, fullerene

1. Introduction

From the past few decades, non-covalent interactions have gained the new heights in the domain of catalysis and supported to construct functional systems of fundamental significance. Infact, the application of non-covalent interactions in the field of catalysis are nowadays studied under a separate branch of science known as supramolecular catalysis [1]. By virtue of non-covalent supramolecular interactions, not only the catalytic efficiency of the existed catalysts has been improved but the novel organocatalysts have also been developed [2]. Amongst various supramolecular interactions, the cation-π and anion-π interactions are most promising in the field of catalysis. In cation-π interactions, the cation interacts with π-basic aromatic systems possessing negative quadrupole moment ($Qzz < 0$), whileas in case of recently recognized anion-π interactions, the anion interacts with π-acidic aromatic systems possessing positive quadrupole moment ($Qzz > 0$). For example, benzene ($Qzz = -9B$) is π-basic in nature which makes it suitable to interact with cations through cation-π interactions (**Figure 1**) [3]. However, the interaction of anion with this system looks counterintuitive, as it will lead to repulsion between anion and π-electron cloud of benzene. For this purpose, researchers have

 IntechOpen

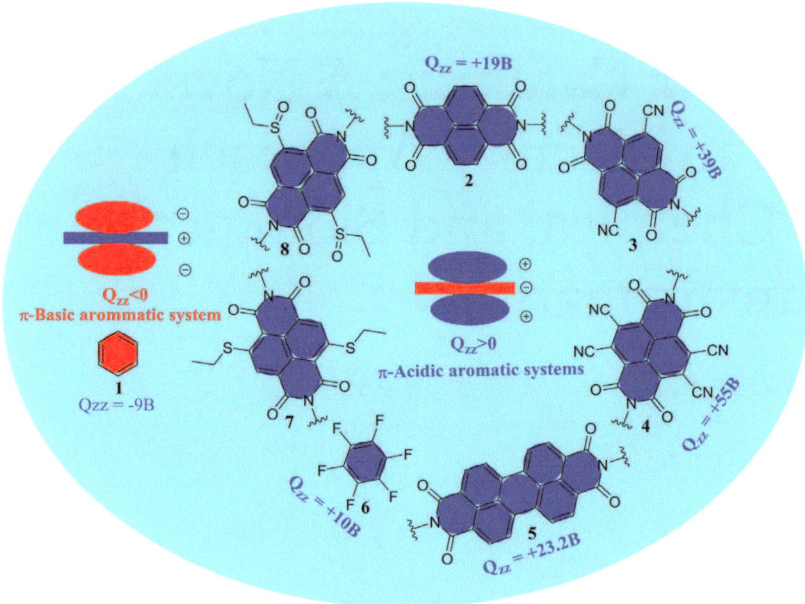

Figure 1.
Structures of diverse π-acidic aromatic systems.

inverted the intrinsic negative Qzz of aromatic systems into positive one (Qzz > 0) by attaching strong electron withdrawing substituents on aromatic systems. By virtue of this, they have generated various π-acidic aromatic scaffolds possessing strong positive quadrupole moment, which in turn interact with diverse anions through anion-π interactions (**Figure 1**) [4].

The recognition of anions is of paramount importance as anions are abundant in nature and play very essential biological role through participation in enzymatic reactions. The transport of anions across biomembranes during different biochemical events makes their recognition even more important [5]. Scientists around the globe are constructing diverse artificial anion receptors, which mimic the function of biosystems and involve anion-π interactions besides other non-covalent interactions in the recognition phenomenon of anions [6]. From the recent theoretical, computational, and experimental investigation, it has been observed that anion-π interactions have shown a promising role in supramolecular catalysis and has given rise to a new concept of anion-π catalysis. The emerging field of anion-π catalysis has not yet much explored in chemistry and until now only few reports are available in literature [7–9]. This is because of the fact that anion-π interactions have recently got experimental evidence, and also there is a dearth of π-acidic aromatic systems being the supreme prerequisite of these interactions [10]. Theoretical and experimental studies have revealed that anion-π catalysis works on the fundamental principle of the stabilization of anionic transition state on π-acidic aromatic surfaces. This stabilization in turn lowers the activation barrier of a particular reaction and hence leads to the formation of a selective desired product quickly under normal reaction conditions [11, 12]. The first evidence of anion-π catalysis came from the Matile's group after carrying out the transmembrane transport of anion by virtue of anion-π interactions [13]. Researchers have developed various anion-π catalysts by adapting different synthetic methodologies [7–9]. It is not feasible herein to discuss such methodologies, but for the convenience of the readers, we have assembled a group of anion-π catalysts used in this chapter (**Figure 2**) [11, 12]. In this chapter, we will discuss the role of these anion-π catalysts in various chemical reactions like

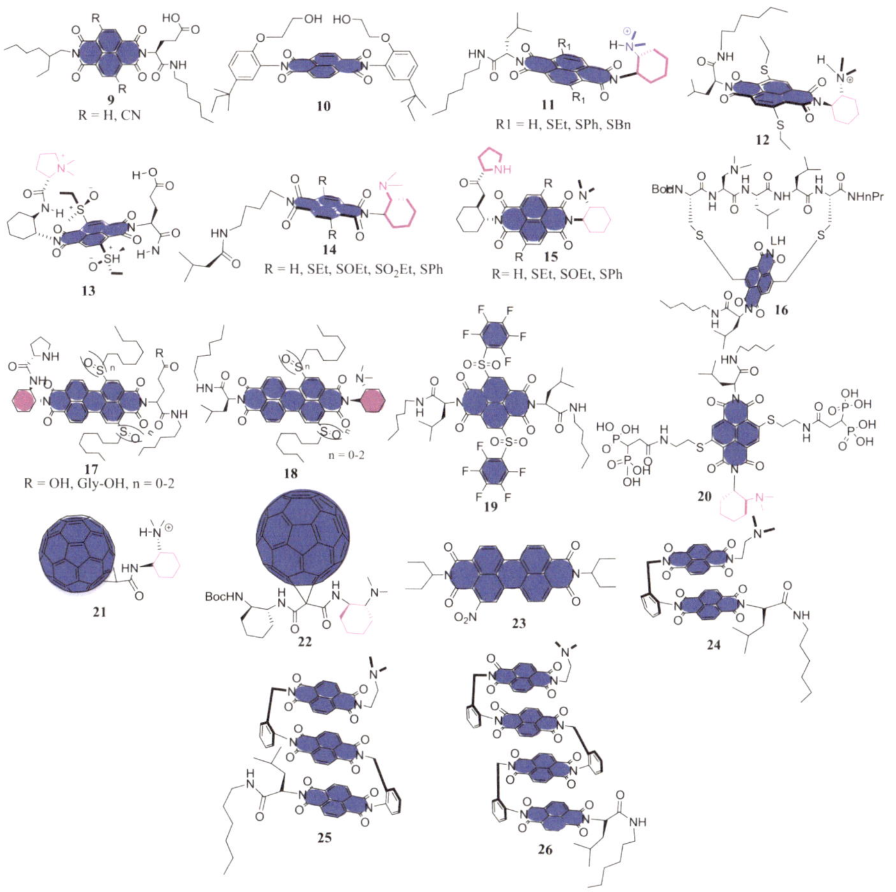

Figure 2.
Structures of various anion-π catalysts used in this chapter.

Kemp elimination, Michael-addition reactions, Diels-Alder reactions, and epoxide ring opening reactions followed by ether and polyether cascade cyclization reactions. Mostly these reactions involve the generation of chiral centers and hence are of prime importance in the arena of asymmetric synthesis.

2. Kemp elimination: a classical tool for anion-π catalysis

The Kemp elimination is a well-known reaction, which involves the abstraction of a proton from the carbon of the benzisoxazole substrate with the help of a catalytic amount of base. This reaction plays an essential biological role and has been documented as an ideal conventional tool for anion-π catalysis by Matile's group. They have carried out this reaction by virtue of the NDI-based anion-π catalysts possessing covalently linked carboxylate base and a solubilizer (alkyl tail) on the π-acidic surface [11]. There occurs the formation of phenolate in the anionic transition-state (**30**) after the proton is abstracted by a covalently attached carboxylate base of NDI based anion-π catalyst (**9**) from the carbon of the substrate (**27**). The anionic transition-state (**30**) acquires stability on the π-acidic surface of catalyst (**9**) by means of anion-π interactions with a magnitude of $\Delta\Delta GTS = 31.8 \pm 0.4$ kJ mol^{-1} along with a catalytic ability of 3.8×10^5 M^{-1} and the transition state recognition (K_{TS}) = 2.7 ± 0.5 μM (**Figure 3**). In order to circumvent the inhibition of the desired

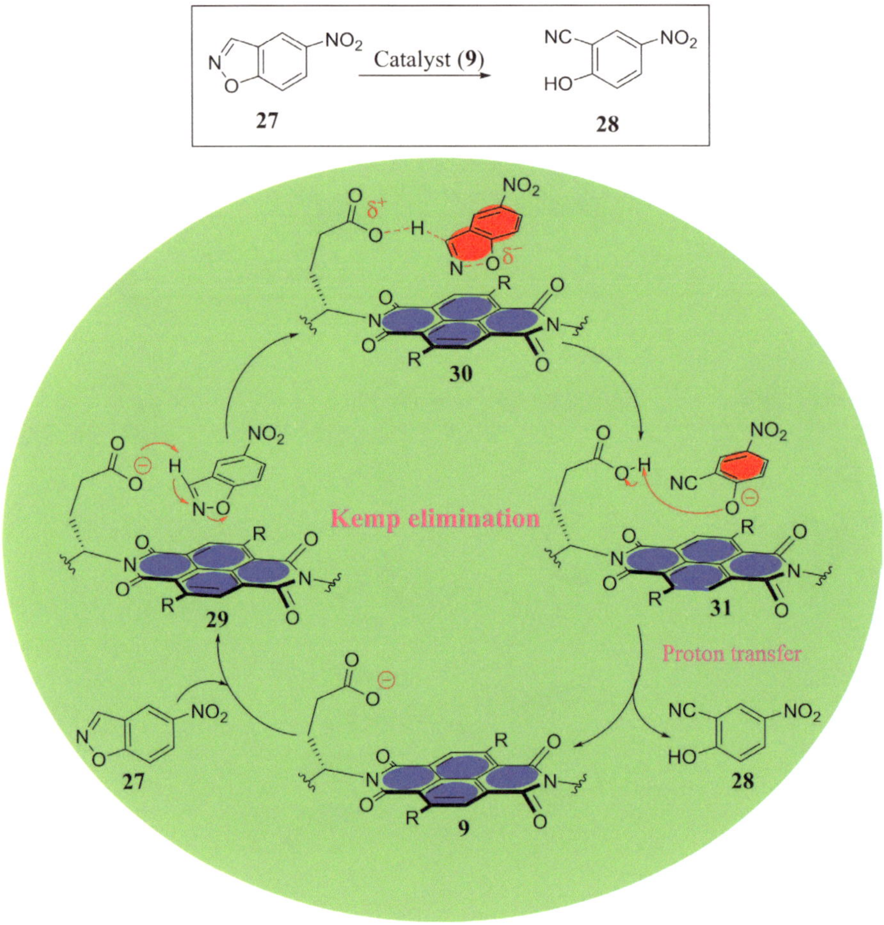

Figure 3.
Schematic illustration of Kemp elimination reaction along with a mechanistic cyclic pathway. Blue color depicts electron deficient and red color implies electron rich.

product, reactive intermediate (**31**) involves protonation of phenolate to yield the desired product **28** with the regeneration of the catalyst (**9**) (**Figure 3**) [11].

3. Michael-addition reactions through anion-π catalysis

Michael-addition, a powerful tool in organic synthesis is a nucleophilic addition reaction which involves the addition of a nucleophile to an α,β-unsaturated carbonyl compounds. It is considered as one of the significant atom economic methods to generate enantioselective and diastereoselective C-C bonds under mild reaction conditions. Knowing the fact that enolates play a vital role in both biology as well as chemistry, Matile's group has revealed the role of anion-π catalysis in enolate chemistry. They have computed the catalytic strength of enolate-π interactions by reporting numerous stereoselective Michael addition reactions on NDI based anion-π catalysts (**Figure 4**) [14–17]. From the ^{1}H-NMR studies, a substantial upfield shift of malonate protons has been observed in cases where the malonate group is covalently linked to the π-acidic NDI surface as compared to protons of unbound diethyl malonate group. This in turn leads to enhanced acidity of the associated malonate group by means of anion-π interactions. In fact, this upfield shift display the location of malonate α-protons above the π-acidic surface [15].

On the other hand, experimental studies have revealed that Michael-addition between malonic acid half thioester (**37**) and enolate acceptor **38** fails without anion-π catalyst and hence leads to solely the formation of a decarboxylation product **40**. Fortunately, the presence of anion-π catalyst for the same reaction leads to the formation of an addition product (**39**) selectively in comparison to decarboxylation product (**40**) (**Figure 5**) [16]. Besides these significant catalytic studies of anion-π interactions, the anion-π catalytic domain has now been explored to iminium, [18] enamine, [15] oxocarbenium, and transamination chemistry [19]. Since the Michael addition reactions through enolate chemistry has a significant relevance to several chemical and biological phenomenon, the outcomes of the studies carried out by Matile and teammates can take the domain of organocatalysis to a next level.

By virtue of positive molecular electrostatic potential (MEP), the fullerene (C_{60}) is considered as a potential candidate for anion-π interactions and has recently been introduced in the field of anion-π catalysis. The exceptional selectivity of fullerenes is actually due to localized positive potential areas termed as π-holes (**Figure 6**). By virtue of purest π-system, fullerenes offer a promising role in the exploration of polarizibility significance in a solution-phase anion-π catalysis [20]. Studies based on these facts have revealed a significant improvement in the selective addition of **37** to **38** by means of a fullerene based catalytic dyad **22** as well as catalytic triad **44** (**Figure 6**). Significantly higher selectivity was achieved after intercalating the methyl viologen **45** in between two fullerenes moieties of a catalytic triad **44** [21, 22]. Besides, the addition product selectivity over the decarboxylation product, conjugate addition protonation reaction of **46** and **47** generating 1,3-non-adjacent stereocenters have been also been carried out on the surfaces of anion-π catalyst **21** composed of fullerene (**Figure 6**) [23].

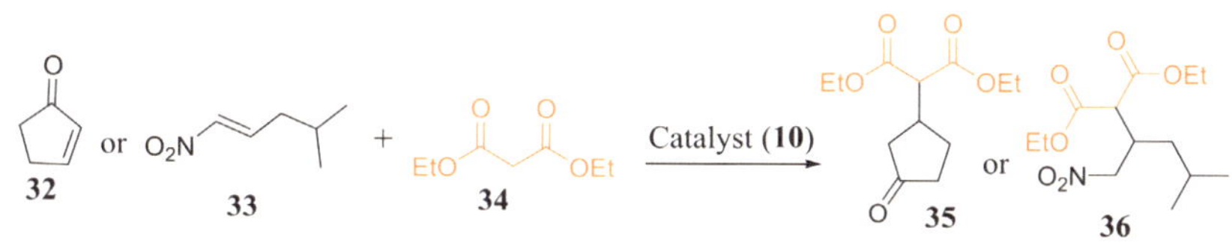

Figure 4.
Schematic representation of Michael addition reaction of **34** *with Michael acceptors (***32** *and* **33***) on the π-acidic NDI surface of* **10**.

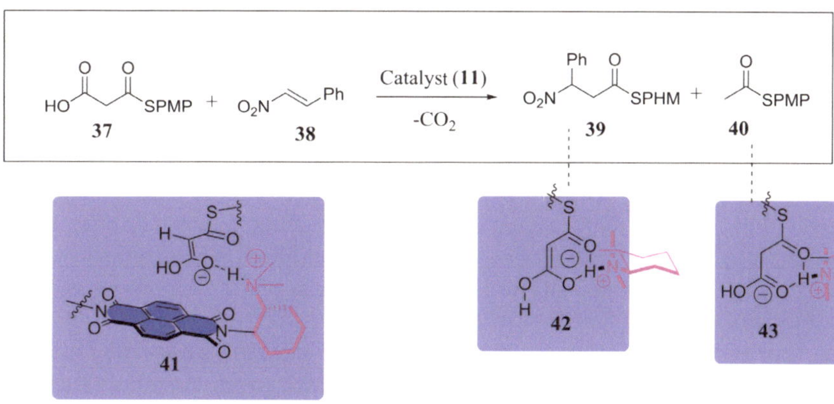

Figure 5.
*Addition (***39***) and decarboxylation (***40***) products in terms of Michael-addition between* **37** *and* **38** *catalyzed by NDI-based anion-π catalyst* **11** *(PMP = p-methoxyphenyl). The molecular structures of transition state (***41***) and of reactive intermediates (***42** *and* **43***) are also given.*

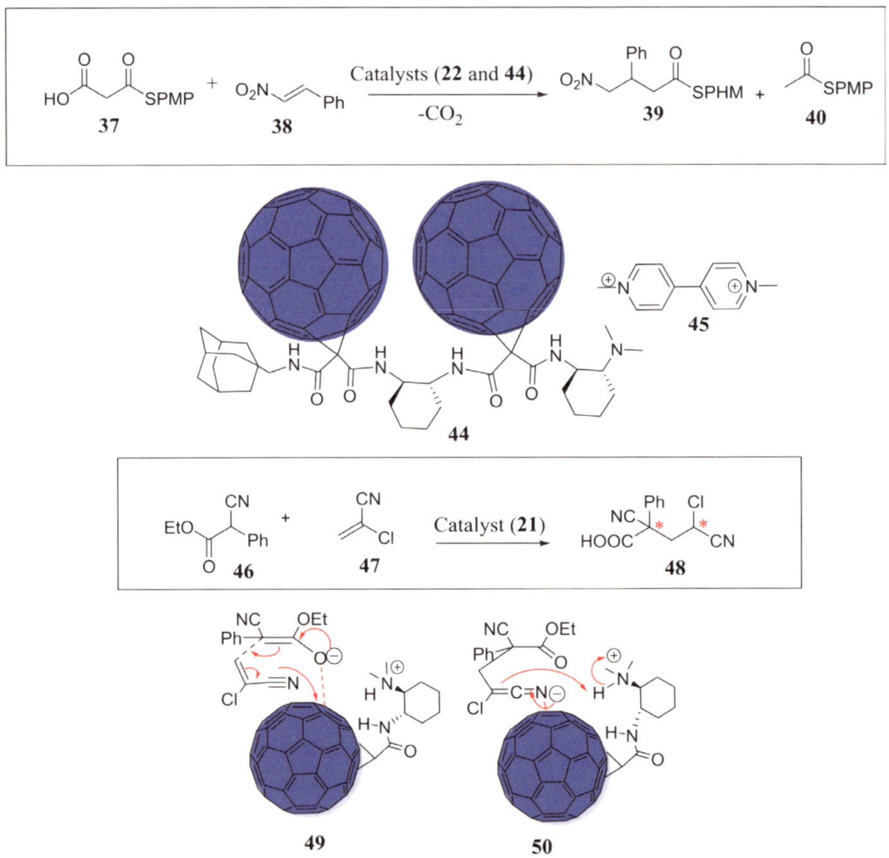

Figure 6.
Schematic illustration of addition (39) and decarboxylation product (40) on anion-π catalytic surfaces of fullerenes (22 and 44) along with the depiction of conjugate addition protonation reaction and anionic transition state stabilization (49 and 50)

Currently, catalysis by means of an electric field has gained a significant interest in molecular transformations, stereoselectivities, and multistep organic synthesis [24]. Electric fields and potentials besides accelerating the reactions have also been shown to activate the conventional catalysts, enzymes, and catalytic pores [25]. Recent studies have revealed that electric fields can just function as a remote control for anion-π catalysts. It has been observed that immobilization of anion-π catalysts is important before applying the electric field to polarize the π-acidic aromatic systems into an induced dipole (μz), which in turn enhance the stabilization of anionic transition state and intermediate on the catalytic polarizable π-surface (**Figure 7**). In order to fulfill these expectations, a bifunctional anion-π catalyst (**20**) has been designed and synthesized by Matile and coworkers for addition product (**39**) selectivity over decarboxylation product (**40**). In their experimental studies, they have attached two diphosphonate groups in the NDI core of **20** through sulphide substituents in order to immobilize the bifunctional catalyst (**20**) on the surface of indium tin oxide (ITO) (**Figure 7**) [26].

In another event, Matile and teammates have used foldamers (**24–26**) for the Michael addition product selectivity over decarboxylated product and noticed that the selectivity gets increased by increasing the length of stacks, thereby displaying anion-$(\pi)_n$-π catalysis (**Figure 8**) [27]. This is actually due to increase in the electronic communication upon increasing the stack length of foldamers. The catalytic activity dependence on electron-sharing has been found to be superlinear on raising

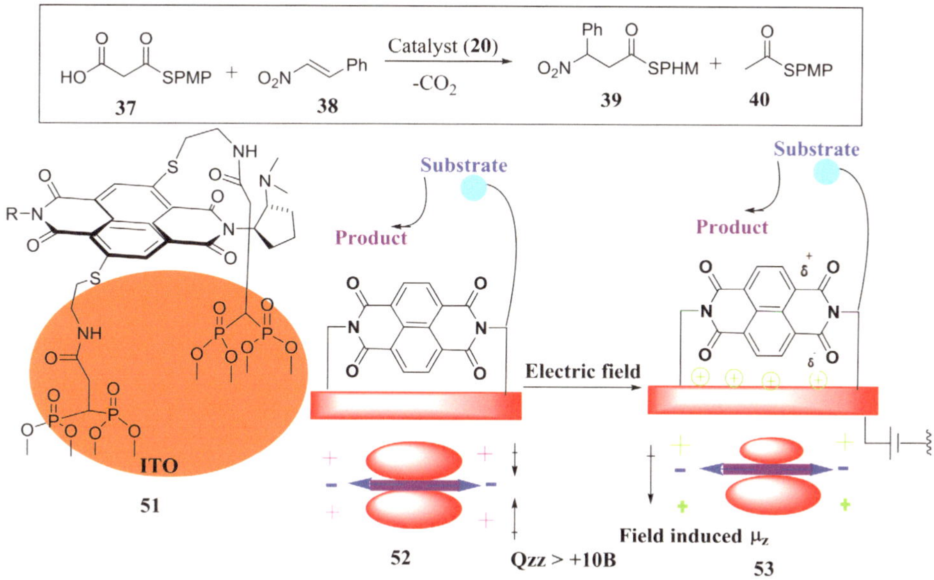

Figure 7.
*Schematic illustration of addition product (39) and decarboxylation product (40) along with immobilized anion-π catalyst **51** and induced dipole (μz).*

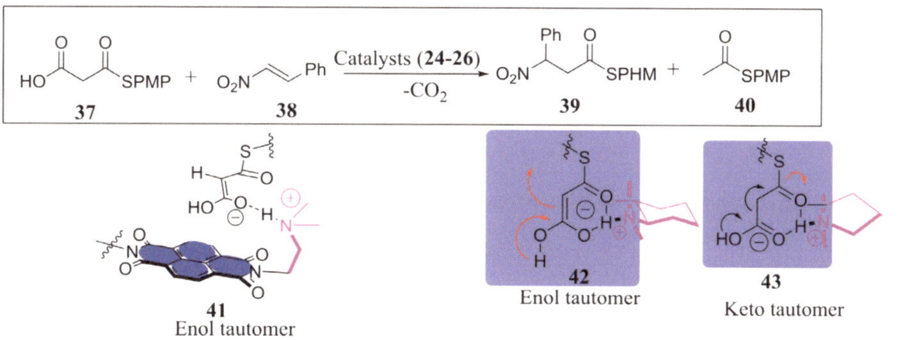

Figure 8.
Schematic depiction of addition product (39) selectivity over decarboxylated product (40) by means π-stacked foldamer assisted anion-$(\pi)_n$-π catalysis.

the stack length. It thus violates sublinear power laws of oligomeric chemistry and reveals the catalytic activity of synergistic amplification over the complete stack [27].

Moreover, quite recently Matile's group has stapled short peptides to NDI-based anion-π catalysts (**16**) in order to generate selective Michael addition product (**39**) over decarboxylative product (**40**) (**Figure 9**). These results regarding anion-π catalysis will serve as an appropriate starting material subsequent to peptides in order to be in operation in larger protein structures and development of anion-π enzymes [28].

Anion-π catalysis play a significant role in the asymmetric synthesis and leads to the generation of chiral isomers selectively. In this regard, the same group has also incorporated NDI moiety in between a carboxylate base and a proline unit for the construction of an anion-π catalyst (**13**). On the π-acidic surface of **13**, they have carried out asymmetric addition of **54** to nitroolefin enamine acceptor **55**. The NDI π-acidic surface helps in the stabilization of transition state of anion near nitronate intermediate by anion-π interactions as can be observed from the structures of transition states (**59** and **60**) (**Figure 10**). By means of the presence of carboxylate base and proline unit at opposite sides of NDI, it has been found that both the rate of enamine addition and enantioselectivity gets enhanced on increasing the π-acidity of NDI [29, 30]. Thus, it gives an essential indication of the participation of anion-π contacts in stereoselectivity. Moreover, anion-π catalyst (**12**) based on NDI was used for the imine isomerization of undesired achiral substrate **57** to the desired chiral product **58** (**Figure 10**) [18].

On the other occasion, the same group has also carried out asymmetric synthesis on anion-π catalytic surfaces of perylenediimides (PDIs). It has been observed that twist in the π-acidic surface determines the catalytic activity of these PDI-based anion-π catalysts in case of Michael addition reactions of enolates and enamines. This is in contrary to the catalytic activity of NDIs, where reducibility of π-surfaces plays a prominent role. Experimental studies have revealed asymmetric addition of **62** to **38** through PDI-based anion-π catalyst (**17**), which leads to the formation of product **63** containing two chiral centers (**Figure 11**) [31]. Furthermore, the PDI-based anion-π catalyst (**18**) also offers Michael addition product (**39**) selectivity over decarboxylated product (**40**) (**Figure 11**).

In another event, Matile's group has also observed anion-π interactions in anion-π enzymes after preparing anion-π enzyme artificially. They have equipped a range of anion-π catalysts with a water-soluble vitamin known as biotin in order to determine the selectivity of Michael addition product (**39**) over decarboxylation one (**40**) in the chemistry of enolates. [32] Additionally, they screened artificially prepared

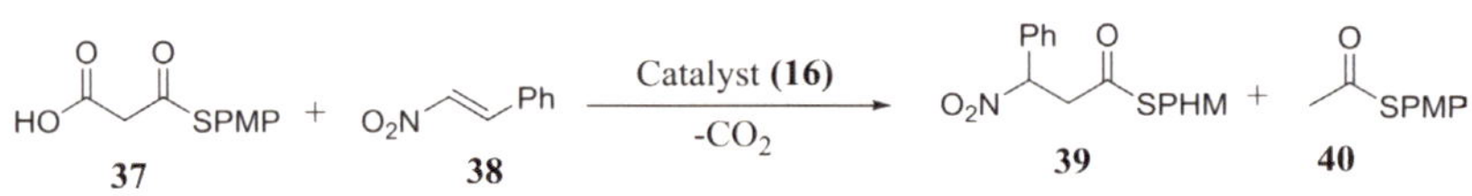

Figure 9.
Schematic depiction of addition product (39) selectivity over decarboxylated product (40) by means of peptide stapled NDI-based anion-π catalyst (16).

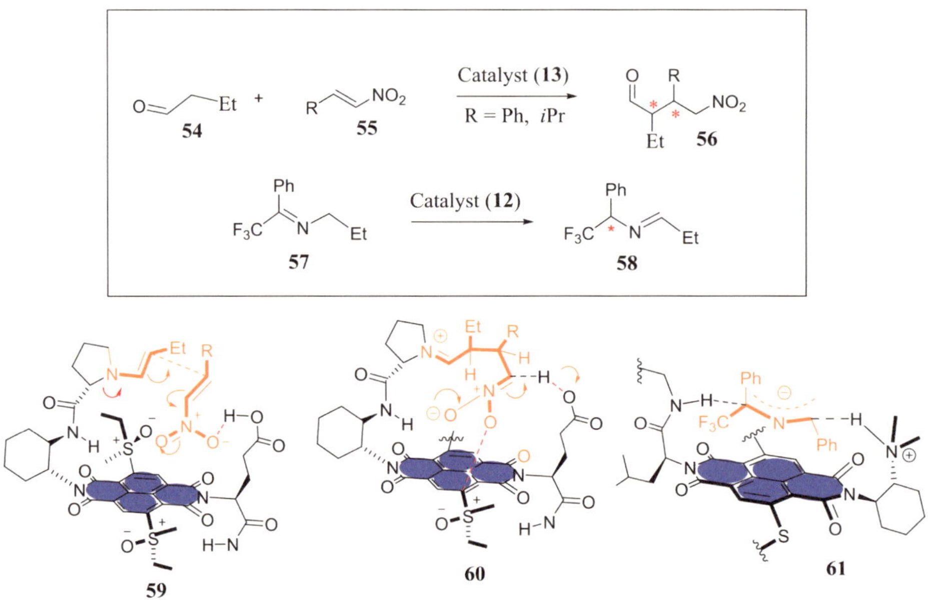

Figure 10.
Representation of asymmetric addition (56) and imine isomerization product (58) along with the molecular structures of transition states (59–61).

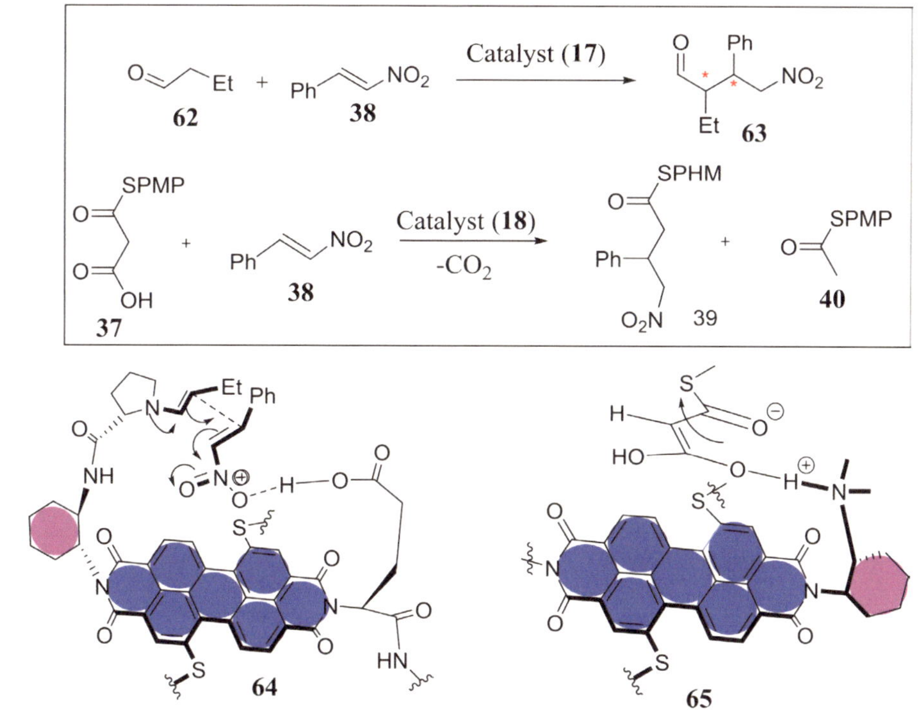

Figure 11.
Asymmetric synthesis of **63** *through PDI-based anion-π catalyst (***17***) and transition state (***64***) depicting formation of C-C bond. Moreover, Michael addition of* **37** *to* **38** *through PDI-based anion-π catalyst (***18***) and a structure of involved transition state (***65***) is also given.*

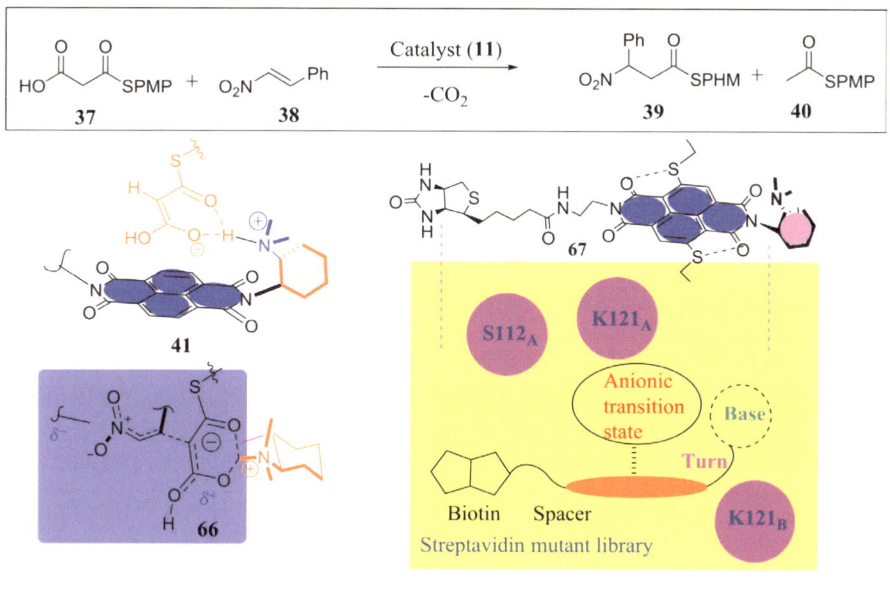

Figure 12.
*Schematic representation of addition (***39***) and decarboxylation product (***40***) by means of anion-π catalyst (***11***). The structures of transition states and reactive intermediates along with the diagrammatic illustration of anion-π enzymes are also given.*

anion-π enzyme against a cluster of the mutants of streptavidin (**Figure 12**). The presence of S112Y mutant leads to desired Michael addition product (**39**) with 95% enantiomeric excess (ee) along with a complete suppression of the decarboxylation

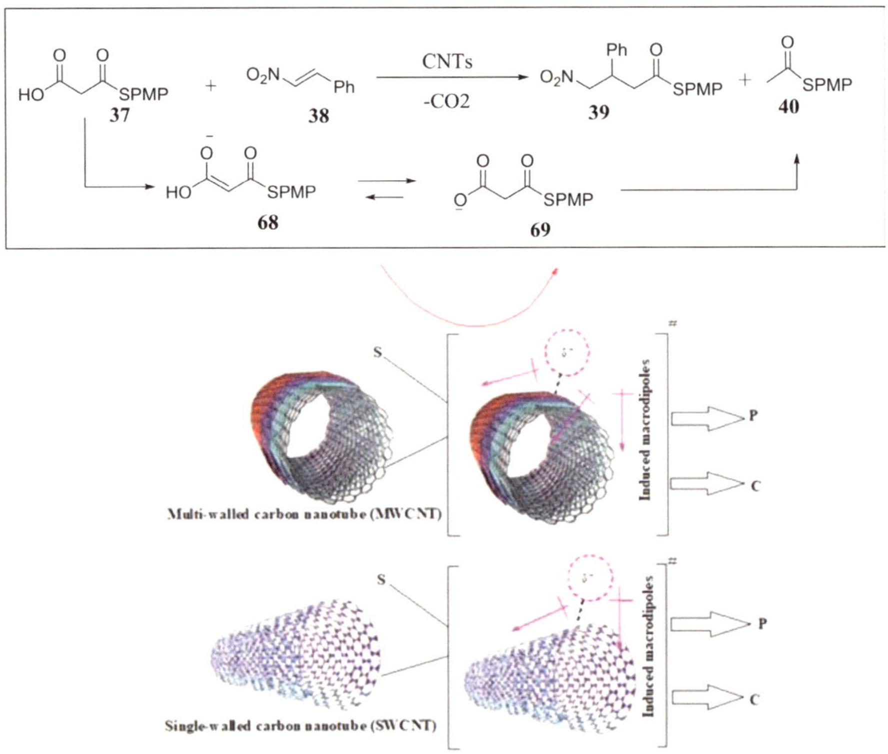

Figure 13.
Schematic illustration of selective addition product (39) on the polarizability induced π-acidic carbon nanotubes (SWCNT and MWCNT).

product (**40**). The existence of anion-π contacts in proteins has been established through the nitrate inhibition of mutant S112Y. The optimum performance has been found at acidic pH = 3, which clearly indicates that enolate gets formed by virtue of the stabilization on π-acidic surfaces. Moreover, K121 mutant has been found in concurrence with the docking results as far as the function of catalyst composed of tertiary amine is concerned at an ideal pH 3. By means of diverse mutants, it is established that enhancing enantioselectivity continuously agrees with the stabilization of particular transition state [32].

More interestingly, the same group has reported innovative anion-π catalysis on the surfaces of carbon nanotubes and synthesized selective addition products on their π-acidic surfaces (**Figure 13**). Studies have revealed that tertiary amine based multi-walled carbon-nanotubes (MWCNT) display much higher efficiency as compared to single-walled carbon nanotubes (SWCNT). This is by virtue of the fact that between and along the nanotubes of MWCNT, there exists a polarizibility induced π-acidic surfaces [33].

4. Anion-π catalysis in action for Diels-Alder reactions

The Diels-Alder reaction discovered in 1928 (Noble prize 1950), a pericyclic [4 + 2] cycloaddition reaction unites diene and dienophile in an atom economic way to yield corresponding Diels-Alder adducts in a regio- and stereoselective manner. Interestingly, this reaction has been used for the synthesis of a plethora of medicinal as well as other compounds. With these thoughts in mind, Matile's group in recent years has successfully carried out Diels-Alder reactions by means of anion-π catalysts

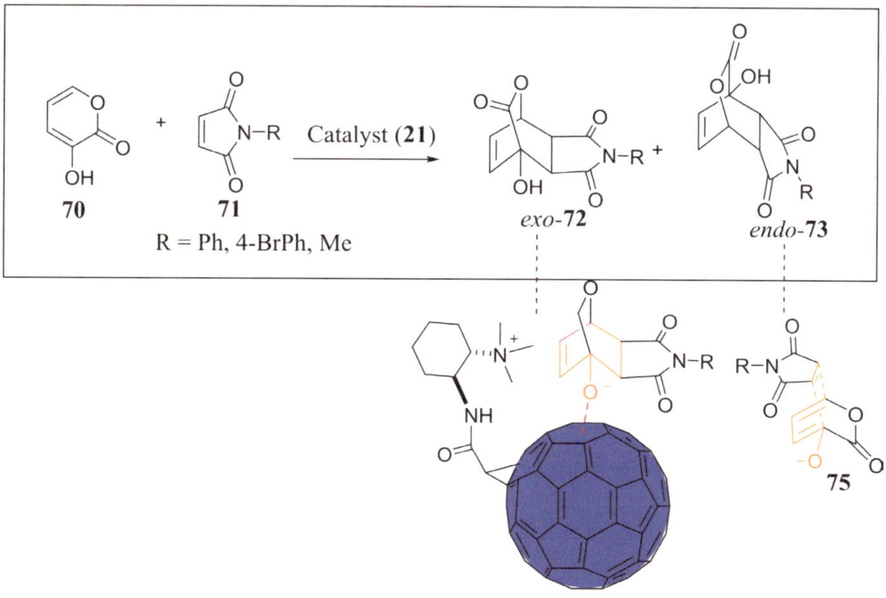

Figure 14.
Diels-Alder reaction between hydroxypyrone (70) and maleimides (71) by means of anion-π catalyst (21). Moreover, the transition states of exo-74 and endo-75 compounds are also given.

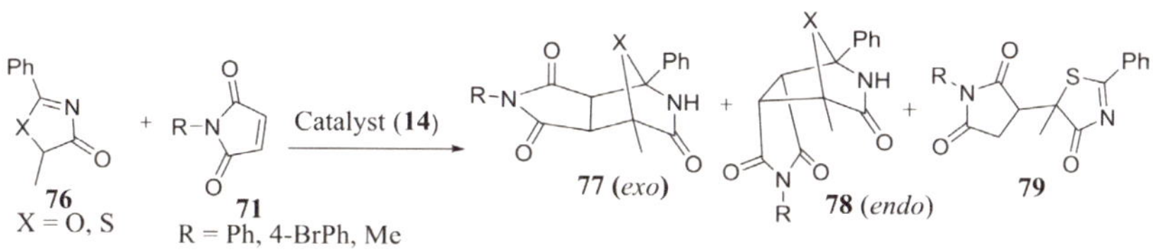

Figure 15.
Michael product (77) catalyzed by triethylaluminium and exo-Diels-Alder product (79) catalyzed by anion-π catalyst (14).

based on fullerenes [34]. During the experimental studies, they have got thermodynamically more stable *exo*-products as compared to generally formed *endo*-products by virtue of the *exo*-transition state stabilization as shown in **Figure 14.**

There is no doubt that the main objective of anion-π catalysis is to discover the reactions of indefinite reactivities and the Diels-Alder reactions of anionic nature offers a first indication in this direction. Matile's group has revealed that that the reaction between **76** and **71** catalyzed by triethylaluminium yields Michael adduct **77** as a major compound along with a Diels-Alder side product **78**. Nevertheless, a reaction of concerted cycloaddition nature by means of anion-π catalyst (14) solely offers *exo*-Diels-Alder compound (**79**) (**Figure 15**) [34].

5. Cascade reactions through anion-π catalysis

Cascade reactions are also known as domino or tandem reactions and comprises of at least two simultaneous consecutive reactions. Herein, the preceding reaction develops a chemical functionality on which a subsequent reaction occurs. Such reactions are of vital importance in the synthesis of complex natural products possessing various chiral centers [35]. During these cascade cyclization reactions, charge displacements are stretched over longer distances. Matile's group has revealed that anion-π catalysis in terms of anion-π interactions is highly capable in the stabilization of these charge displacements. With the help of anion-π catalysis, the cascade reactions on the π-acidic catalytic surface leads to generation of bicyclic asymmetric products possessing four chiral centers (**Figure 16**). Moreover, cascade cyclization reactions through anion-π catalysis are also in action in the generation of asymmetric cyclohexane moieties containing five chiral centers generated in a single step on π-acidic catalytic surface (**Figure 17**). The concept of anion-π catalysis also play a central role in other cascade reactions on the π-acidic aromatic surface. For instance, the reaction of **93** with **94** takes place in a cascade way on the π-acidic aromatic surface (**Figure 18**). The greatest outcome of anion-π catalysis in cascade cyclization reactions is anion-π cinchona fusion catalyst [36].

Epoxide ring opening followed by ether and polyether cascade cyclic reactions are considered as conventional reactions in chemical and biological sciences. Nowadays, these reactions are also considered as attractive tools for anion-π catalysis. To this context, Matile's group has reported some functional systems

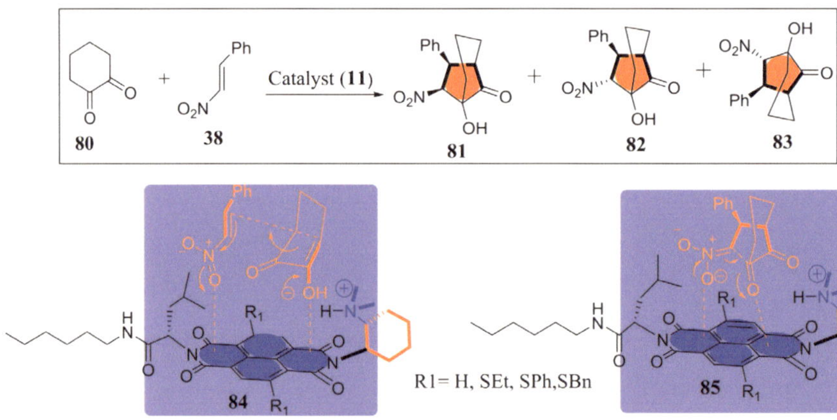

Figure 16.
Schematic representation of the generation of anion-π catalyzed bicyclic cascade products and the structures of involved anion-π transition states.

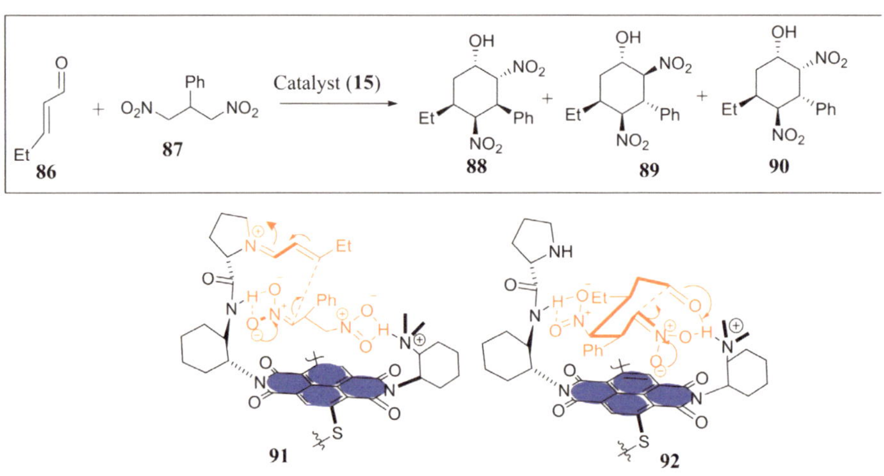

Figure 17.
Schematic illustration of asymmetric anion-π catalyzed cyclohexane rings through cascade cyclisation reaction. Moreover, the structures of anion-π transition states are also given.

Figure 18.
*Schematic depiction of anion-π catalyzed cascade reaction of benzaldehyde derivative (*93*) with ethyl acetoacetate (*94*).*

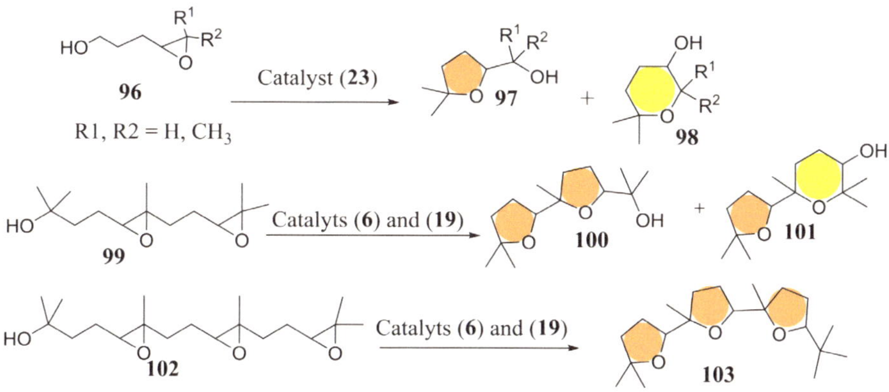

Figure 19.
*Schematic illustration of epoxide ring opening followed by ether and polyether cascade cyclization reactions by virtue of anion-π catalysts (*6, 19, *and* 23*).*

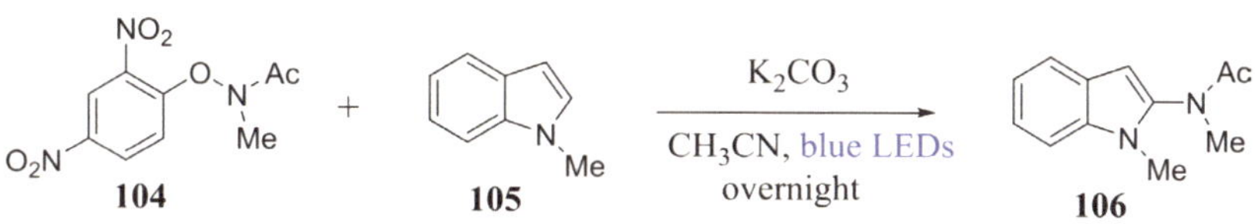

Figure 20.
Photoamidation reaction between 104 *and* 105 *assisted by anion-π interactions.*

which operate through anion-π interactions and show autocatalysis. Studies have revealed that aromatic π-acidic surfaces involve epoxide ring opening followed by ether cyclization without any activating group (**Figure 19**) [37]. Quite recently, they have also observed exceptional high autocatalysis on the π-acidic surfaces of hexafluorobenzene and substituted NDI's as far as epoxide ring opening followed by cyclisation reactions are concerned. This unique characteristic of autocatalysis not only adds complexity to reaction mechanisms but also offers intriguing perspectives towards future developments [38, 39].

Besides the above-mentioned catalytic relevances of anion-π interactions in the domain of catalysis, amidation reactions driven by light have also been carried out by means of these interactions. It has been observed that anion-π interactions helps in the stabilization of transient complex formed between electron deficient moiety **104** and carbonate or phosphate anion. This complex in turn undergoes cleavage of N-O bond in the presence of light to offer amidyl radical, which is later trapped by heteroaromatic system (**105**) to offer the desired product **106** (**Figure 20**) [40].

6. Conclusions and outlook

Anion-π catalysis in general operates on the fundamental principle of anionic transition state stabilization on π-acidic aromatic surfaces and offers a novel approach towards diverse molecular transformations. Over the past seven years, steady advancement has been made in the domain of anion-π catalysis with regard to the design of catalyst and the scope of the reaction. Considering the significance of polarizability, it is believed that there will be the emergence of more hidden occurrences of immature anion-π catalysis in the near future. The unconventional anion-π catalysis gains an optimistic outlook from the immense impact of current developments made with conventional cation-π and ion-pairing interactions. It is thus expected that anion-π catalysis will eventually offer new mechanisms and access to new reactivities. However so far, anion-π catalysis fails in the general expectation to produce novel products. Nevertheless, efforts are being carried out all across the globe to meet the general expectations of anion-π catalysis to offer access to novel products with exceptional features, which are far outside the scope of conventional catalysis.

Acknowledgements

We are grateful to DST-SERB New Delhi for financial support (Project File no. ECR/2017/000821). I. A. R. thanks CSIR, New Delhi for the award of the JRF and SRF. R. A. thanks Jamia Millia Islamia, New Delhi for providing the necessary research facilities.

Conflicts of interest

The authors declare no conflicts of interest.

Author details

Ishfaq Ahmad Rather and Rashid Ali*
Organic and Supramolecular Functional Materials Research Laboratory, Department of Chemistry, Jamia Millia Islamia, New Delhi, India

*Address all correspondence to: rali1@jmi.ac.in

References

[1] P. Ballester, A. Scarso, Editorial: Supramolecular aspects in catalysis, Front. Chem. 2019;7.174. doi:10.3389/fchem.2019.00174.

[2] J.B. Baruah, Principles and advances in supramolecular catalysis, CRC Press, Boca Raton : CRC Press, Taylor & Francis Group, 2019. doi:10.1201/9780429059063.

[3] D.A. Dougherty, The cation-π interaction, Acc. Chem. Res. 2013;46:885-893. doi:10.1021/ar300265y.

[4] I.A. Rather, S.A. Wagay, R. Ali, Emergence of anion-π interactions: The land of opportunity in supramolecular chemistry and beyond, Coord. Chem. Rev. 2020;415:213327. doi:10.1016/j.ccr.2020.213327.

[5] N. Busschaert, C. Caltagirone, W. Van Rossom, P.A. Gale, Applications of supramolecular anion recognition, Chem. Rev. 2015;115:8038-8155. doi: 10.1021/acs.chemrev.5b00099

[6] I.A. Rather, S.A. Wagay, M.S. Hasnain, R. Ali, New dimensions in calix[4]pyrrole: the land of opportunity in supramolecular chemistry, RSC Adv. 2019;9:38309-38344. doi: 10.1039/C9RA07399J

[7] Y. Zhao, Y. Domoto, E. Orentas, C. Beuchat, D. Emery, J. Mareda, N. Sakai, S. Matile, Catalysis with anion-π interactions, Angew. Chemie Int. Ed. 2013;52:9940-9943. doi:10.1002/anie.201305356.

[8] T. Lu, S.E. Wheeler, Quantifying the role of anion-π interactions in anion-π catalysis, Org. Lett. 2014;16:3268-3271. doi:10.1021/ol501283u.

[9] C. Wang, S. Matile, Anion-π catalysts with axial chirality, Chem.- A Eur. J. 2017;23:11955-11960. doi:10.1002/chem.201702672.

[10] M. Giese, M. Albrecht, K. Rissanen, Experimental investigation of anion-π interactions – applications and biochemical relevance, Chem. Commun. 2016;52:1778-1795. doi:10.1039/C5CC09072E.

[11] Y. Zhao, C. Beuchat, Y. Domoto, J. Gajewy, A. Wilson, J. Mareda, N. Sakai, S. Matile, Anion-π catalysis, J. Am. Chem. Soc. 2014; 136:2101-2111. doi:10.1021/ja412290r.

[12] Y. Zhao, Y. Cotelle, L. Liu, J. Lopez-Andarias, A.-B. Bornhof, M. Akamatsu, N. Sakai, S. Matile, The emergence of anion-π catalysis, Acc. Chem. Res. 2018;51:2255-2263. doi:10.1021/acs.accounts.8b00223.

[13] J. Mareda, S. Matile, Anion-π slides for transmembrane transport, Chem. - A Eur. J. 2009;15:28-37. doi:10.1002/chem.200801643.

[14] Y. Cotelle, S. Benz, A.-J. Avestro, T.R. Ward, N. Sakai, S. Matile, Anion-π catalysis of enolate chemistry: Rigidified Leonard turns as a general motif to run reactions on aromatic surfaces, Angew. Chem. 2016;128: 4347-4351. doi:10.1002/ange.201600831.

[15] Y. Zhao, N. Sakai, S. Matile, Enolate chemistry with anion-π interactions, Nat. Commun. 2014;5:3911. doi:10.1038/ncomms4911.

[16] Y. Zhao, S. Benz, N. Sakai, S. Matile, Selective acceleration of disfavored enolate addition reactions by anion-π interactions, Chem. Sci. 2015;6:6219-6223. doi:10.1039/C5SC02563J.

[17] F.N. Miros, Y. Zhao, G. Sargsyan, M. Pupier, C. Besnard, C. Beuchat, J. Mareda, N. Sakai, S. Matile, Enolate stabilization by anion-π interactions: Deuterium exchange in malonate dilactones on π-acidic surfaces,

Chem. - A Eur. J. 2016;22:2648-2657. doi:10.1002/chem.201504008.

[18] L. Liu, Y. Cotelle, A.-J. Avestro, N. Sakai, S. Matile, Asymmetric anion-π catalysis of iminium/nitroaldol cascades to form cyclohexane rings with five stereogenic centers directly on π-acidic surfaces, J. Am. Chem. Soc. 2016;138:7876-7879. doi:10.1021/jacs.6b04936.

[19] L. Liu, S. Matile, Anion-π transaminase mimics, Supramol. Chem. 2017;29:702-706. doi:10.1080/10610278.2016.1258118.

[20] D.S. Sabirov, Polarizability as a landmark property for fullerene chemistry and materials science, RSC Adv. 2014;4:44996-45028. doi:10.1039/C4RA06116K.

[21] J. Lopez-Andarias, A. Frontera, S. Matile, Anion-π catalysis on fullerenes, J. Am. Chem. Soc. 2017;139:13296-13299. doi:10.1021/jacs.7b08113.

[22] J. Lopez-Andarias, A. Bauza, N. Sakai, A. Frontera, S. Matile, Remote control of anion-π catalysis on fullerene-centered catalytic triads, Angew. Chemie Int. Ed. 2018;57:10883-10887. doi:10.1002/anie.201804092.

[23] X. Zhang, L. Liu, J. Lopez-Andarias, C. Wang, N. Sakai, S. Matile, Anion-π catalysis: focus on nonadjacent stereocenters, Helv. Chim. Acta. 2018;101:e1700288. doi:10.1002/hlca.201700288.

[24] S. Shaik, R. Ramanan, D. Danovich, D. Mandal, Structure and reactivity/selectivity control by oriented-external electric fields, Chem. Soc. Rev. 2018;47:5125-5145. doi:10.1039/C8CS00354H.

[25] S. Shaik, D. Mandal, R. Ramanan, Oriented electric fields as future smart reagents in chemistry, Nat. Chem. 2016;8:1091-1098. doi:10.1038/nchem.2651.

[26] M. Akamatsu, N. Sakai, S. Matile, Electric-field-assisted anion-π catalysis, J. Am. Chem. Soc. 2017;139:6558-6561. doi:10.1021/jacs.7b02421.

[27] A.-B. Bornhof, A. Bauza, A. Aster, M. Pupier, A. Frontera, E. Vauthey, N. Sakai, S. Matile, Synergistic anion-$(\pi)_n$-π catalysis on π-stacked foldamers, J. Am. Chem. Soc. 2018;40:4884-4892. doi:10.1021/jacs.8b00809.

[28] A. Pham, S. Matile, Peptide stapling with anion-π catalysts, Chem. – An Asian J. 2020;15:1562-1566. doi:10.1002/asia.202000309.

[29] Y. Zhao, Y. Cotelle, A.-J. Avestro, N. Sakai, S. Matile, Asymmetric anion-π catalysis: enamine addition to nitroolefins on π-acidic surfaces, J. Am. Chem. Soc. 2015; 137:11582-11585. doi:10.1021/jacs.5b07382.

[30] M. Akamatsu, S. Matile, Expanded chiral surfaces for asymmetric anion-π catalysis, Synlett. 2016;27:1041-1046. doi:10.1055/s-0035-1561383.

[31] C. Wang, F.N. Miros, J. Mareda, N. Sakai, S. Matile, Asymmetric anion-π catalysis on perylenediimides, Angew. Chemie Int. Ed. 2016;55:14422-14426. doi:10.1002/anie.201608842.

[32] Y. Cotelle, V. Lebrun, N. Sakai, T.R. Ward, S. Matile, Anion-π enzymes, ACS Cent. Sci. 2016;2:388-393. doi:10.1021/acscentsci.6b00097.

[33] A. Bornhof, M. Vazquez-Nakagawa, L. Rodriguez-Perez, M. Angeles Herranz, N. Sakai, N. Martin, S. Matile, J. Lopez-Andarias, Anion-π catalysis on carbon nanotubes, Angew. Chem. Int. Ed. 2019;58:1-5. doi:10.1002/ange.201909540.

[34] L. Liu, Y. Cotelle, A.-B. Bornhof, C. Besnard, N. Sakai, S. Matile, Anion-π

catalysis of Diels-Alder reactions, Angew. Chemie Int. Ed. 2017;56:13066-13069. doi:10.1002/anie.201707730.

[35] I. Vilotijevic, T.F. Jamison, Epoxide-opening cascades in the synthesis of polycyclic polyether natural products, Angew. Chemie Int. Ed. 2009;48:5250-5281. doi:10.1002/anie.200900600.

[36] L. Liu, Y. Cotelle, J. Klehr, N. Sakai, T.R. Ward, S. Matile, Anion-π catalysis: bicyclic products with four contiguous stereogenic centers from otherwise elusive diastereospecific domino reactions on π-acidic surfaces, Chem. Sci. 2017;8:3770-3774. doi:10.1039/C7SC00525C.

[37] X. Zhang, X. Hao, L. Liu, A.-T. Pham, J. Lopez-Andarias, A. Frontera, N. Sakai, S. Matile, Primary anion-π catalysis and autocatalysis, J. Am. Chem. Soc. 2018;140:17867-17871. doi:10.1021/jacs.8b11788.

[38] M. Paraja, X. Hao, S. Matile, Polyether natural product inspired cascade cyclizations: autocatalysis on π-acidic aromatic surfaces, Angew. Chemie Int. Ed. 59 (2020) 15093-15097. doi:10.1002/anie.202000681.

[39] M. Paraja, S. Matile, Primary anion–π catalysis of epoxide-opening ether cyclization into rings of different sizes: Access to new reactivity, Angew. Chemie Int. Ed. 2020;59:6273-6277. doi:10.1002/anie.202000579.

[40] L. Buglioni, M.M. Mastandrea, A. Frontera, M.A. Pericas, Anion-π interactions in light-induced reactions: Role in the amidation of heteroaromatic systems with activated N -aryloxyamides, Chem. – A Eur. J. 2019;25:11785-11790. doi:10.1002/chem.201903055.

Chapter 7

Chiroptical Polymer Functionalized by Chiral Nanofibrillar Network

Hirotaka Ihara, Makoto Takafuji and Yutaka Kuwahara

Abstract

Chirality is one of the basic factors that influence a wide range of activities from chemical synthesis to tissue construction in life phenomena. Recently, researchers have attempted to use chirality as an optical signal. In animals, it is used to transmit information to insects and crustaceans, and it has also been confirmed that it promotes growth in plants. This chapter presents a new organic system that produces a chiral optical signal, that is, circularly polarized luminescence (CPL), which has been attracting attention in recent years. In particular, the chapter is focused on the generating CPL through chirality induction with the chiral self-assembling phenomenon and explaining its application as an optical film.

Keywords: self-assembling, supramolecular gel, nanofibril, circular dichroism, circularly polarized luminescence

1. Introduction

Fiber materials have applications in various industrial fields. One of these is compounding with bulk polymer materials to improve and strengthen their physical strength by controlling factors such as the bending elastic modulus, tensile strength, and thermal expansion coefficient. Generally, the higher the dispersion of fibers in polymer, the higher the effect on polymer.

Typical examples of fibers with industrial applications are carbon fibers, aramid fibers, and whiskers. In recent years, cellulose nanofibers [1] having a nano-sized diameter have attracted attention not only because they can be manufactured using inexhaustible and inexpensive raw materials but also because the strength of each cellulose nanofiber is 1/5th that of iron and 5 times the strength [2, 3]. Therefore, if the process cost and the method of dispersion of cellulose nanofibers in bulk materials can be improved and established, it will be used in various materials and fields, such as automobiles and home appliances.

In this chapter, we focus on the low light-scattering property due to the sufficiently small diameters as optical materials because transparency is one of the most important physical elements. Inorganic glass is a transparent material with excellent heat resistance, light resistance, and chemical resistance. Therefore, it is almost universally used as a partition plate during material conversion and energy conversion using light as an energy source. However, inorganic glass is heavy, inflexible, and fragile, and to improve impact resistance, laminated glass with an organic polymer as an interlayer film is often used. Therefore, there is a need for

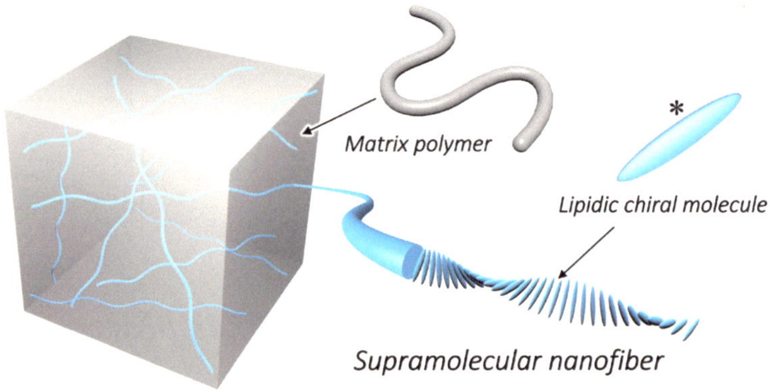

Figure 1.
A strategic approach to highly functionalizing polymer materials through self-assembled growth into nanofibers.

a transparent material that is light, soft, and has good processability. Transparent polymers are suitable to replace glass because organic polymers are essentially light at the elemental level and can be adjusted for hardness and flexibility. In addition, it is attractive that expansive functionality can be tailor-made at the monomer level.

One of the methods for imparting functions while taking advantage of such characteristics of organic polymers is to blend fillers. This method is particularly applicable to general-purpose polymers and is widely used for function enhancement and function conversion. Among them, nano-sized fillers are attracting particular attention in applications that require transparency because they have low light-scattering properties. However, there are several challenges in the use of nanofillers, and at present, reducing process costs and simplifying the compounding process are the main focus issues. In particular, simplification of the compounding process is a universal problem that results from the high specific surface area of nanomaterials and is thus often a barrier to development and research. This problem also exists in the cellulose nanofibers mentioned above.

This chapter outlines an approach (**Figure 1**) that utilizes nanofiber formation [4–6] using the self-assembly of small molecules as a new strategic method for functionalizing polymer materials; also, an example of its application as an optical material is introduced.

2. Opt functional nanofibers by self-assembly

While various methods for creating nanofibers have been developed, it is difficult to unravel a mass of nanofibers once entangled by a process that is complicated and requires high energy. A typical example is cellulose nanofibers (CNFs). CNFs are environmentally friendly and have excellent potential functions; therefore, ensuring efficient dispersion of CNFs is essential to develop their applications. The method introduced in this chapter is not used for unraveling entangled nanofibers, but rather a for growing nanofibers in a bulk polymer (**Figure 1**). Therefore, because the material used is a low-molecular-weight organic substance, a nanofiber-forming material that is structurally compatible with the polymer material can be selected. This approach promises a surfactant-free process in compounding polymeric materials.

To grow nanofibers in polymers, we utilize self-assembling gelation, which is a novel method in solution systems. Well-known molecular materials for this purpose are cholesterol derivatives [7], sugar derivatives [8], and amino acid derivatives [9, 10] (**Figure 2**). These materials associate in a self-organizing manner by promoting weak

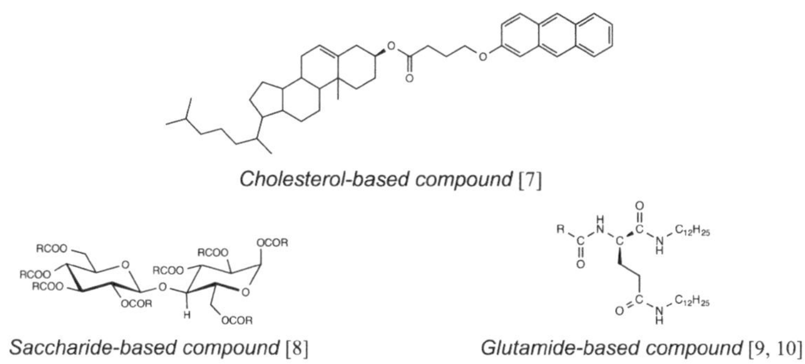

Figure 2.
Compounds that cause a molecular gelation phenomenon due to the formation of nanofiber-like aggregates by self-assembling [7–10].

intermolecular interactions and orientation states that are attributed to the molecular design. It is also known that the presence of chirality in the molecular structure tends to make the association form a nanofiber-like structure. After a certain concentration, nanofibers become entangled, resulting in gelation of the solution, which is called a molecular gel or supramolecular gel and is different from a typical polymer gel. **Figure 3** summarizes amino acid-derived molecular gel-forming materials [5, 9–76] discussed in this chapter.

2.1 Principle of opt functionalization by molecular gelation

Following are some essential requirements for small molecules to assemble in solution to form nanofibers. (1) The molecule has a part that is dispersible in a solvent, (2) a part that exhibits a relatively strong interaction among molecules, (3) a rigid and non-bulky structure that promotes intermolecular stacking, which is useful as an auxiliary function, and finally, (4) a mechanism that facilitates one-dimensional growth of the association structure set in the molecular structure. The existence of molecular chirality that promotes twisted molecular stacking can be considered the most effective mechanism. A molecular gel comprising a cholesterol derivative, reported by Weiss et al. in 1987, is one of the small molecules that is suitable for this application (**Figure 2**) [7]. However, it has been reported that nanofiber-like aggregates are formed by various low-molecular-weight substances when not limited to the apparent gel state. For example, the authors found that nanofiber-like aggregates could be formed by fibrous bilayer membrane structures in aqueous systems [9, 77]. Since Weiss's report, non-cholesterol derivatives, such as derivatives from amino acid, peptide and polysaccharide, are known as molecular gel-forming substances, and they have led to remarkable developments in this field [78–80].

Figure 3 shows typical amino acid derivatives that produce molecular gels. These molecules satisfy all four requirements described above. These derivatives are more advantageous than cholesteryl derivatives because hydrogen bonds are based on intermolecular interactions, and therefore, by properly tuning the molecular structure, nanofiber structures can be formed in a wide range of solvents, such as water and organic solvents. Further, as shown in the electron micrograph of **Figure 4**, they all form nanofibrillar aggregates although the detailed aggregation morphology differs based on the structure. The formed aggregates have a nano-sized diameter and are rarely bundled, even if they are mixed in the polymer as described later; therefore, there are few whitening due to light scattering. This is an extremely important property for producing a transparent opt functional film.

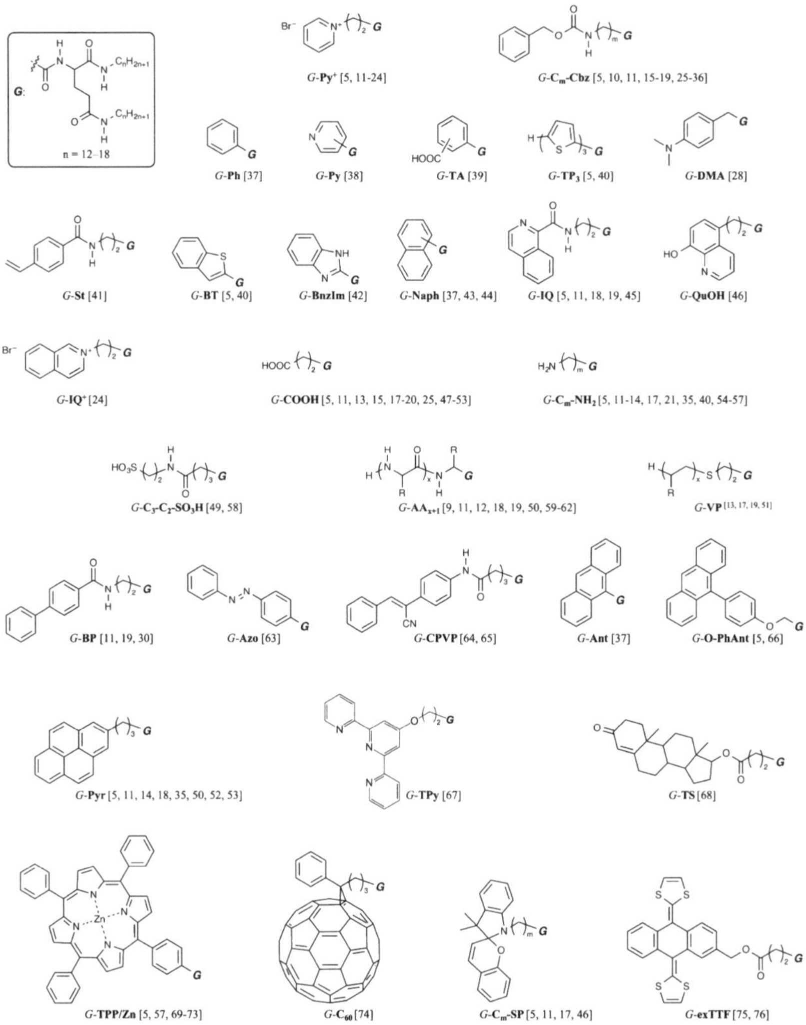

Figure 3.
Examples of amino acid-derived molecular gel-forming materials for nanofibrillar network formation [5, 9–76].

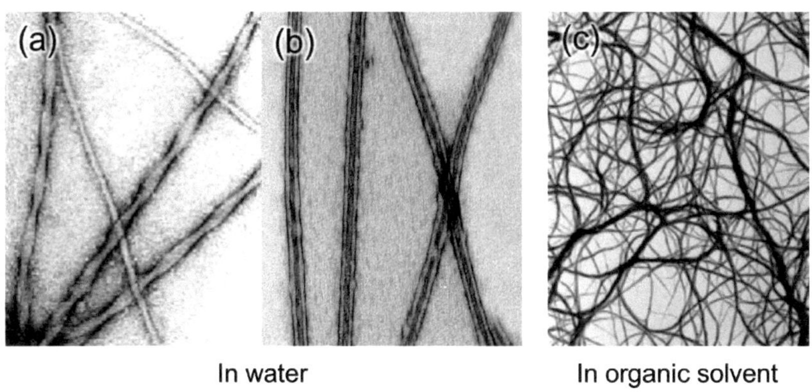

Figure 4.
*Typical examples of nanofibrillar aggregates from amino acid derivatives: (a) and (b) were observed by G-***Py**$^+$ *in aqueous systems, and (c) was observed by G-***Cbz** *in benzene [12, 25, 81, 82].*

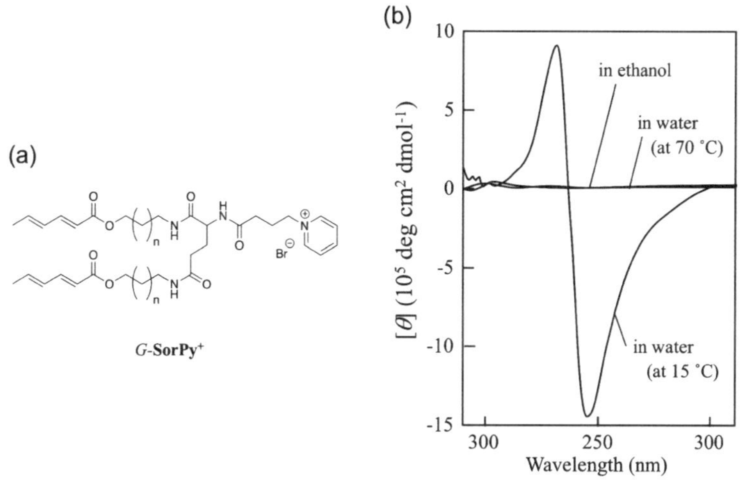

Figure 5.
*Chemical structure (a) and CD spectra (b) of G-***SorPy**⁺ *[12] G-***SorPy**⁺ *is in a highly ordered state at 15 °C in water.*

Molecular gels made from amino acid derivatives often exhibit amplified chirality. Amino acids have an asymmetric center, because of which they act as chiral materials by themselves, and a quite low circular dichroism (CD) intensity, which is a measure of the magnitude of chirality. When these amino acid derivatives are aggregated with hydrogen bonds and a twist in a certain direction occurs in the orientation state, a very large CD signal (Cotton effect) can be obtained; a typical example is shown in **Figure 5**. *G-***SorPy**$^+$ has no asymmetric carbon or optical activity in the sorbyl group. Therefore, the CD signal is not observed in the molecularly dispersed state. On the other hand, when the molecules form an ordered orientation state, a chiral stacking state is created in the sorbyl group (**Figure 5b**), because of which, a very large CD signal is observed around the absorption band of the sorbyl group. Such amplified chirality is often referred to as secondary chirality to distinguish it from chiral-source chirality. That is, the nanofibers produced by self-assembly are likely to exhibit amplified secondary chirality, and this supramolecular function is directly linked to the principle of chiroptical functionalization for polymers, which we will focus on in this section.

2.2 Luminous nanofiber

There are roughly two methods for imparting optical functions to self-assembled nanofibers. One is the single-component system, which is a method for introducing a luminescent functional group during molecular design. The other method is the binary system, which is a method for combining self-assembled nanofibers as a template with a luminophore as a guest molecule by molecular interactions such as electrostatic interaction.

2.2.1 Single-component system for chirality enhancement

First, a molecular gel system introduced using pyrene, which is widely known as an organic fluorescent dye, is described. *G-***Pyr** in **Figure 3** is an amino acid derivative in which pyrene is introduced. Since *G-***Pyr** is a hydrophobic compound, it is insoluble in water but soluble in various organic solvents. When *G-***Pyr** is dissolved in hot cyclohexane or toluene to a concentration of approximately 5 mM and returned to room temperature, it forms a transparent gel [25, 26]. **Figure 6** shows a transparent gel state and a xerogel prepared by freeze drying. Transmission electron microscopy (TEM) and scanning electron microscopy (SEM) indicated that clear gel and white solid were formed from fibrous components. When benzene was added to 5 mM *G*-**Pyr** gel to dilute it to a 0.2 mM mixture, which was heated to form a solution, no gelation was observed even when it was cooled to room temperature. However, the formation of fibrous aggregates was confirmed by TEM observation. These facts conclude that *G*-**Pyr** forms nanofibers, even under low concentration conditions (apparently in a solution state), that do not form a physical gel state.

When a 0.2 mM benzene solution of *G*-**Pyr** was excited at 350 nm, corresponding to the absorption maximum of a pyrene group, a fluorescent spectrum with an emission maximum at 455 nm could be obtained, as shown by the solid line in **Figure 7a**. This is characteristic of the excimer from pyrene. No similar spectrum is observed for the starting material, pyrenebutanoic acid (dotted lines in **Figure 7a**), which can be attributed to the monomeric state. These facts suggest that under the condition that *G*-**Pyr** forms nanofibrillar aggregates, the pyrenyl groups are concentrated from each other, wherein excimers are likely to be formed [26–28].

The CD spectrum also shows the formation of an oriented state by the aggregation of *G*-**Pyr**. As shown in **Figure 7b**, *G*-**Pyr** produced from the L-enantiomer produces an extremely large positive signal (Cotton effect). Because there is no chirality in the

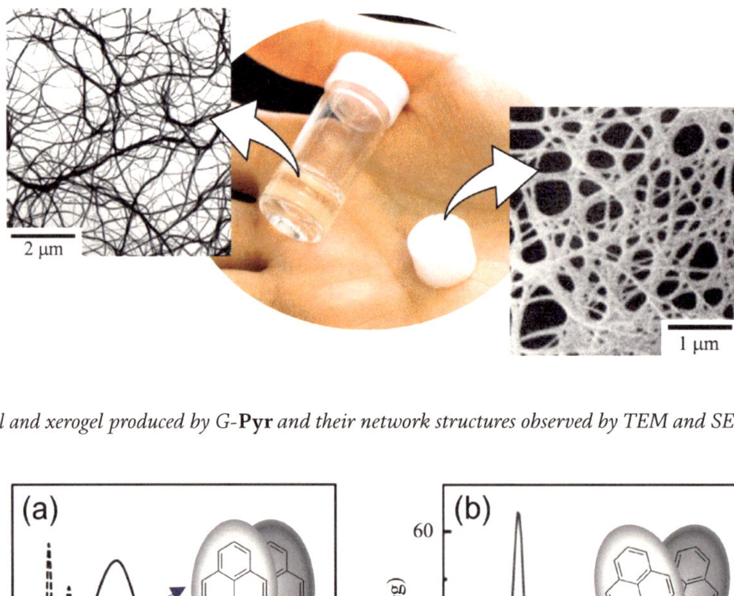

Figure 6.
*Benzene gel and xerogel produced by G-**Pyr** and their network structures observed by TEM and SEM [25].*

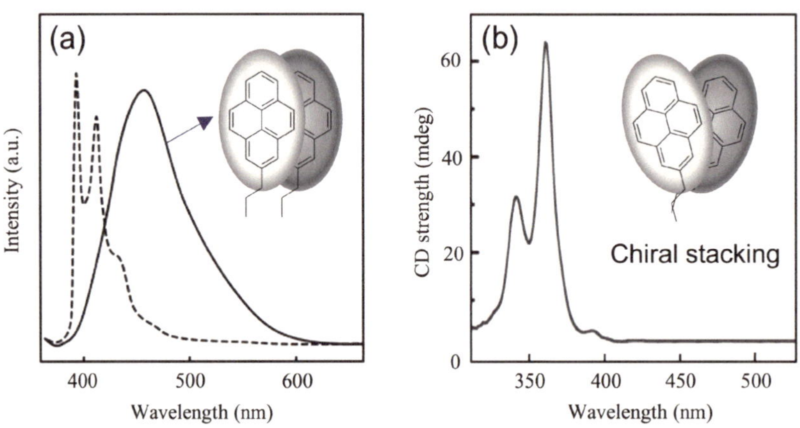

Figure 7.
*Fluorescent (a) and CD (b) spectra of G-**Pyr** solution (0.2 mM in benzene). The dashed line (a) indicates the fluorescent spectrum from pyrenebutanoic acid [26].*

pyrene group itself, this suggests chirality induction from an amino acid moiety to the pyrene group. No similar Cotton effect is observed under conditions where *G*-**Pyr** is in a molecularly dispersed state. To explain this large Cotton effect, it is necessary to consider a phenomenon in which nanofiber-like aggregates are formed by chiral stacking with pyrene groups, as shown in the inset of **Figure 7b**. In *G*-**Pyr**, three amide bonds around the chiral carbon should be focused on as the driving force for producing a chiral stacking state because when the amide bond is changed to an ester bond, a spherical aggregate morphology called a vesicle is preferentially formed over the nanofibrillar aggregates [26].

G-**BT** is an amino acid derivative with a benzothiophene group as a photofunctional group [82]. When *G*-**BT** was dissolved in methylcyclohexane, the fluorescent color changed significantly from yellowish green to bluish purple, depending on the temperature (**Figure 8a**). The confocal laser photomicrograph in **Figure 8c** shows that *G*-**BT** is dispersed in the solution as a fluorescent fibrillar aggregate. Moreover, when observing the fluorescent spectrum, an emission maximum can be observed around 550 nm at room temperature (**Figure 8b**). Since the emission wavelength of the monomer is approximately 330 nm, the Stokes shift differs by nearly 220 nm. This difference is accompanied by that in the fluorescence lifetimes. In fact, it was found that the lifetime in this new emission band around 550 nm was on the order of milliseconds (**Figure 8d**). Presumably, the nanofibrillar aggregates form a stacking state close to the nanocrystals around the benzothiophene group. When heating is used to promote relaxation of this orientation state or trifluoroacetic acid is added as a hydrogen bond inhibitor, the emission around 550 nm disappears and shifts to 330 nm, and the fluorescence lifetime is of the order of nanoseconds, which corresponds to that of the monomeric state of the benzothiophene group.

2.2.2 Binary system for chirality enhancement

While the method of obtaining chirality enhancement by the single-molecule system outlined in Section 2.2.1 has a high degree of freedom in molecular design, there are practical limits in the synthetic process. As a solution to this problem, a binary system has been proposed in which a chiral molecular gel is used as a template material with a highly-ordered microenvironment and is combined with a non-chiral luminescent material. The advantage of this approach is that even a

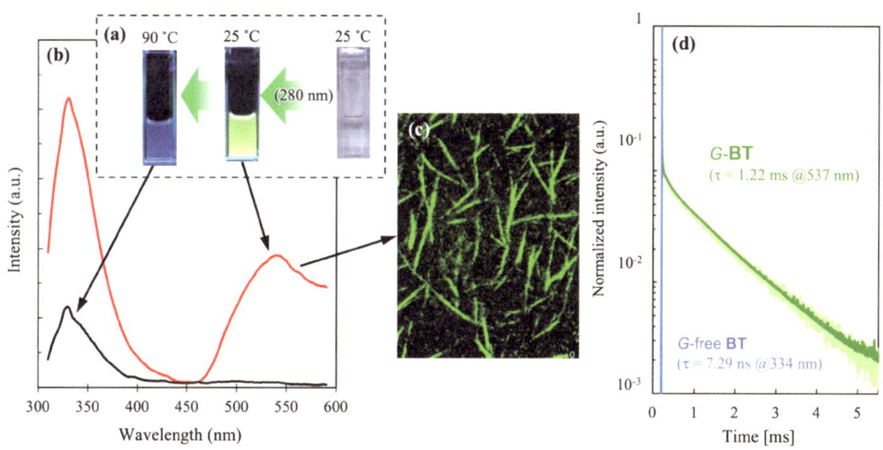

Figure 8.
*Photographs (a), fluorescent spectra (b), confocal image (c), and decay curves (d) of G-**BT** in methylcyclohexane. [83] - reproduced by permission of the Royal Society of Chemistry.*

luminescent dye with a complicated structure has few synthetic chemical restrictions because introducing chirality into it is not necessary.

In this method, the wavelength of the amplified CD signals can be easily controlled based on the dye selected by incorporating an appropriate interaction system between the chiral template orientation and the luminescent dye. This idea has been attempted by several researchers over the years. For example, when a polyamino acid with an ionic functional group in the side chain interacts with a particular dye, dimer formation and the accompanying formation of secondary chirality of the dye are observed [84]. A similar phenomenon is observed in the interaction with polysaccharides that form secondary structures [85]. **Figure 9** shows the CD spectra of hyaluronic acid and cyanine dyes. Because there is no asymmetric carbon in the cyanine dye used, it can be concluded that the CD spectra in **Figure 9** are induced.

There are many examples of chirality induction due to the combination of molecular gels and non-chiral dyes. **Figure 10** shows an example of induced chirality as the binary system. In this case, *G*-**Py**$^{+}$, which has a pyridinium-based cationic group that promotes electrostatic interaction, is adopted as an amino acid-based molecular gel-forming substance, and it is combined with a selected anionic cyanine-based dye [21]. It was clarified that such induced chirality can be observed even for a single atom such as bromine. **Figure 11** shows an example in which induced chirality is expressed by the binding of bromine ions to a chiral molecular gel generated from a gemini surfactant [86].

2.2.3 Stimuli-responsive chirality

Since molecular gelation by a low-molecular-weight compound is a phenomenon induced by molecular assembly, there is a critical gelation concentration, and it

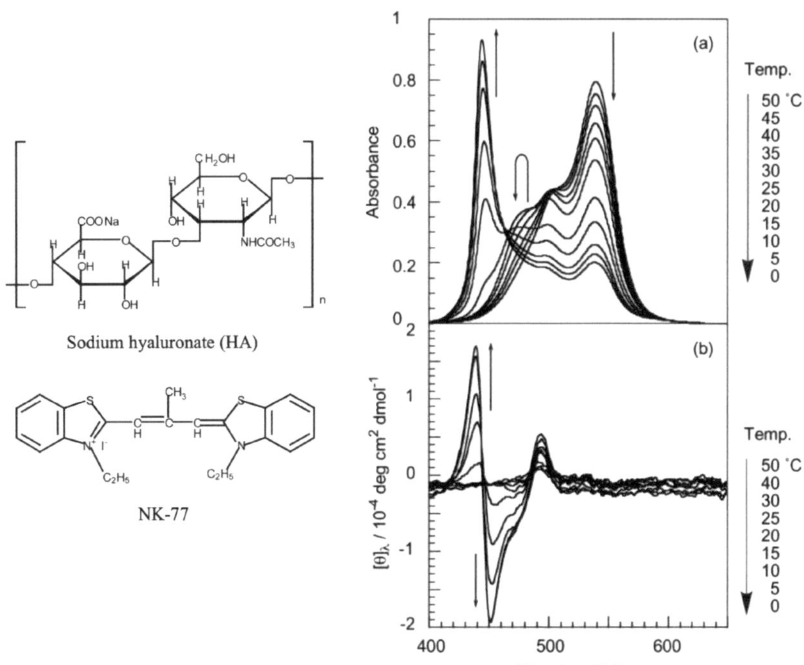

Figure 9.
Example of induced CD to dye from chiral polysaccharide template. (a) Visible absorption and (b) CD spectra of cyanine dye (NK-77) in the presence of sodium hyaluronate (HA) in MeOH–H_2O mixture. [85] - reproduced by permission of the Royal Society of Chemistry.

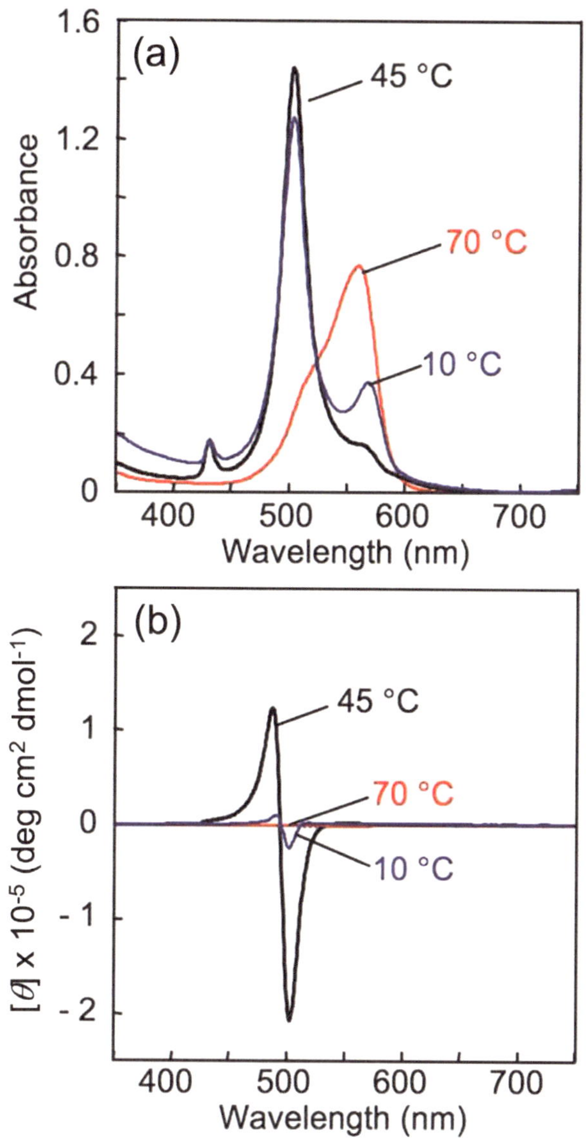

Figure 10.
Example of induced CD to dye from chiral bilayer membrane template. (a) UV–visible absorption and (b) CD spectra of cyanine dye (NK-2012) in the presence of G-$\mathbf{Py}^+$ *in water. [21] - reproduced by permission of Springer Nature.*

varies depending on the kind of solvent. Moreover, the hydrocarbon chain in the molecular structure has a phase transition phenomenon corresponding to the melting point. Therefore, it is easy to understand that molecular gels can have both lyotropic and thermotropic properties.

A typical example of a lyotropic property is given in *G*-$\mathbf{SorPy}^+$ in **Figure 5**; *G*-$\mathbf{SorPy}^+$ exhibits a large CD intensity in water, but not in ethanol [12]. Similar lyotropic behavior can be observed in *G*-**Pyr**. Its fluorescent color in cyclohexane is remarkably different from that in chloroform, and this is attributed to the difference in their solubilities.

An example of thermotropic behavior is the amino acid derivative *G*-$\mathbf{Py}^+$ with a pyridinium group [4, 81]. As shown in **Figure 12a**, the CD intensity varies significantly depending on the temperature. This is because the *G*-$\mathbf{Py}^+$ aggregates

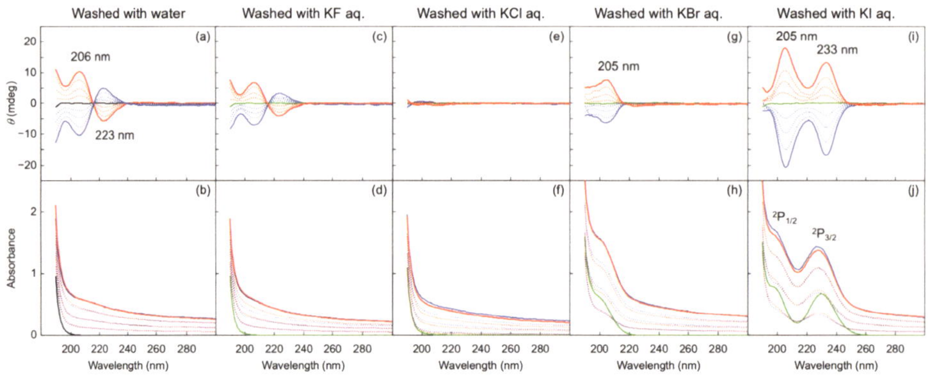

Figure 11.
Example of induced CD to bromine as a single atom from chiral self-assembly-based template. CD and UV–visible absorption spectra of silica-coated L-(red) and D-(blue lines) self-assembled nanohelices in water (0.05–0.20 mg mL^{-1}) obtained by washing. Black and green spectra represent water (b) and 0.05 mM-KX aqueous solutions where X are F (d), Cl (f), Br (h) and I (j). [86] - reproduced by permission of the Royal Society of Chemistry.

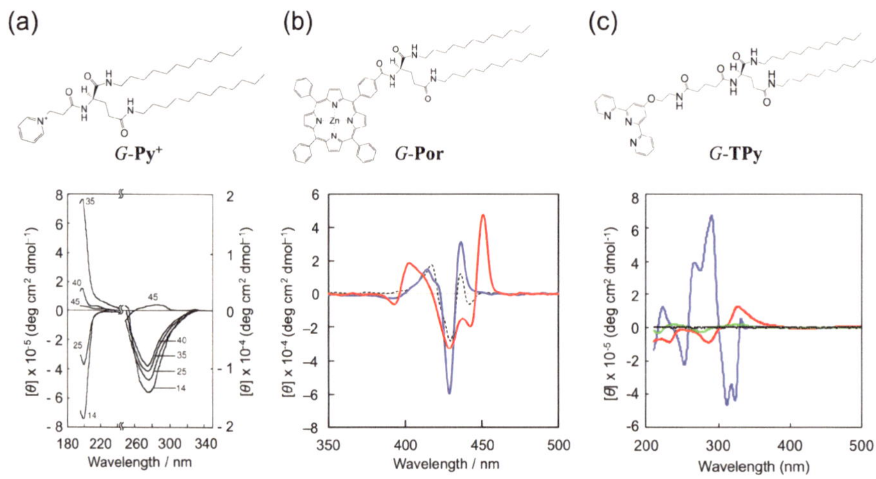

Figure 12.
*Examples of CD responsibilities in G-***Py**$^+$ *(a), [4, 81] G-***por** *(b), [87] and G-***TPy** *(c) [67] assemblies by external factors. [4, 81, 87] - reproduced by permission of the chemical Society of Japan (CSJ). [67] - reproduced by permission of the Royal Society of Chemistry.*

have two phase transition temperatures. Beyond these two-phase transition temperatures, the lateral diffusion rate of the molecule increases even in the aggregated state, and the secondary chirality based on the molecular orientation disappears.

Molecular gels are sensitive not only to solvent types and temperatures but also to other external factors. **Figure 12b** shows the CD spectra of the amino acid derivative *G*-**Por** with a porphyrin as a functional group. In cyclohexane, a very large CD intensity is observed near the absorption band of porphyrin, but since there is no asymmetric carbon or chirality at the porphyrin site, the expression of the large CD signal is attributed to chiral orientation among the porphyrin moieties in cyclohexane gel. The addition of an imidazole derivative to this *G*-**Por** gel remarkably changes the CD pattern, but such a change does not occur with the addition of pyridine at the same concentration, [87] which is attributed to the different coordination abilities of the guest molecules.

Figure 12c shows the CD pattern of a molecular gel with amino acids having a terpyridyl group, *G*-**TPy**. The intensity and pattern of the CD spectra change significantly depending on the type of metal coordinated to the terpyridyl group [67].

As described above, chiral molecular gels can significantly change chirality with respect to various external factors, so their use for sensing using this responsiveness is also attractive.

2.3 Circularly polarized luminescent molecular gel

Circularly polarized luminescence (CPL) is angle-independent and contains light information not found in normal light, and therefore, it is expected to be used in various applications. Potential applicability is found in many industrial fields such as biometric recognition systems, light sources for plant factories, storage memories, and 3D displays. On the other hand, considering organic materials that generate CPL, although there are numerous research results, [88–110] their strength (optical purity) is low and they have not yet been put into practical use.

Figure 13 summarizes the structural formula of single-molecule CPL-generating dyes [90, 92, 94, 96, 99, 103–110]. Most conventional organic materials that generate CPL are chiral fluorescent dyes that utilize the twisting and straining of molecules, [90] and therefore, require complicated chemical synthesis and purification processes. To overcome these problems and fabricate a CPL generation system with higher optical purity, the authors utilized self-assembled nanofiber systems formed from amino acid-derived molecular gels.

As mentioned in Section 2.1, *G*-**Pyr** is a fluorescent substance with chirality; therefore, it is considered as a good candidate, but the resultant CPL intensity (g_{lum} value) was of the order 10^{-3} [66]. It is not superior to the chiral organic fluorescent molecules reported so far. One reason for this has emerged as an inference that there is no optical activity in the excimer emission of pyrene.

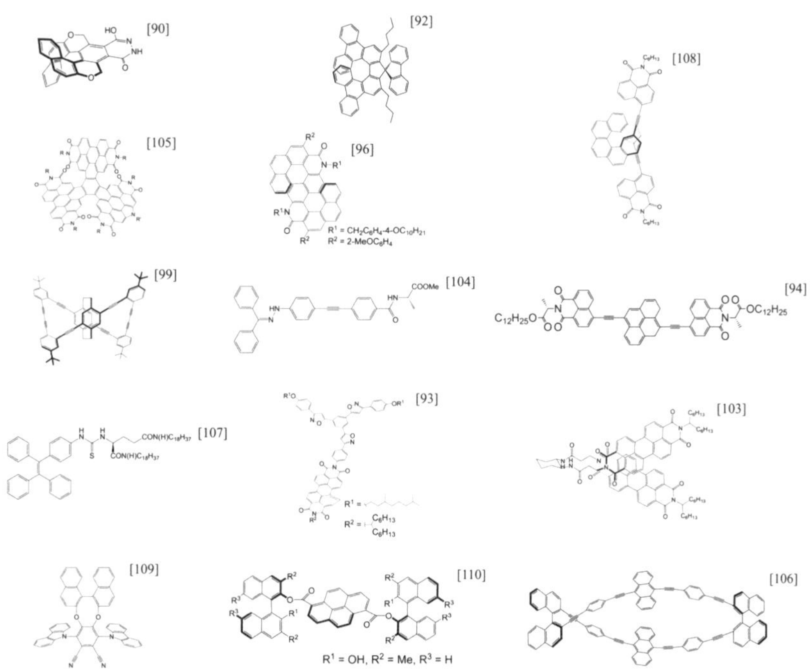

Figure 13.
Examples of CPL-generating fluorescent dyes [90, 92, 94, 96, 99, 103–110].

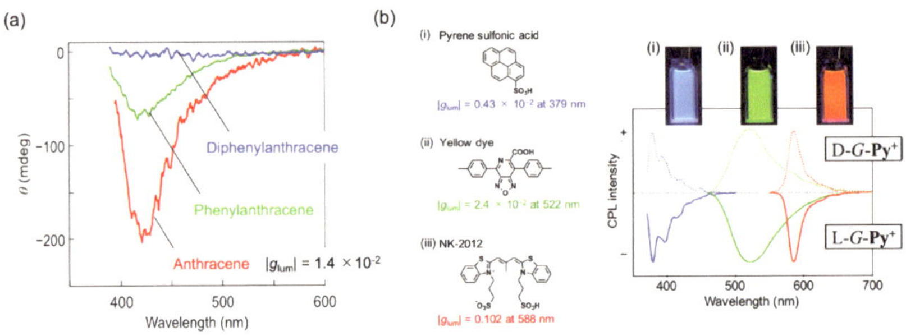

Figure 14.
*Example of induced CPL by a binary system constructed of non-chiral fluorescent dye with chiral molecular gel template. (a) CPL spectra of anthracene, phenylanthracene, and diphenylanthracene in the presence of G-***Py**$^+$ *in aqueous solutions at 2 °C. Excitation wavelength was 340 nm. [Anthracene derivertives] = 12.5 μM. [G-***Py**$^+$*] = 0.5 mM. [81] - reproduced by permission of the Chemical Society of Japan (CSJ). (b) Tunability of CPL emission bands. Blue, green, and red CPL spectra were obtained with (i) pyrenesulfonic acid, (ii) Yellow dye, and (iii) NK-2012, respectively. All the CPL spectra were measured in aqueous systems at 20 °C in the presence of L- and D-enantiomers of G-***Py**$^+$ *with fluorophores. The concentrations of G-***Py**$^+$ *and the fluorophores were 0.5 mM and 12.5 μm, respectively. The excitation wavelengths were 330, 420, and 500 nm in pyrenesulfonic acid, Yellow dye, and NK-2012. All the CPL spectra were normalized at the highest peak of the fluorescence emissions corresponding to the CPL signals. [24] - reproduced by permission of John Wiley and Sons.*

On the basis of this finding, we applied a binary method wherein an existing non-chiral fluorescent dye was incorporated into a molecular gel as a chiral template for CPL generation. This method is advantageous because it is not necessary to introduce chirality into the fluorescent dye, because of which, the degree of freedom in synthetic chemistry is greatly increased. In addition, the light emitting region of the CPL can be easily tuned by proper selection of the dye. As a result, the highest value in CPL ($|g_{lum}| > 0.1$) was obtained using a molecular gel from *G*-**Py**$^+$ as a chiral template with a hydrophobic fluorescent dye such as anthracene (**Figure 14a**) [81] and an anionic cyanine-based fluorescent dye (**Figure 14b**) [24] combined. **Figure 14b** also shows that the emission wavelength can be easily tuned by selecting pyrene sulfonic acid or yellow dye as the fluorescent dye and that the positive and negative can be reversed by the enantiomers of the template [24].

The other example of CPL generation is presented by **Figure 15** [111]. In this case, the *G*-**COOH**-based molecular gel having a carboxyl group is used as a chiral template, considering solubility in a non-aqueous solvent. It was confirmed that the induced CD and the induced CPL were expressed in the binary system combined with the cationic dye. However, in this case, it was also confirmed that the coexistence of a base such as triethylamine in the solution significantly improved

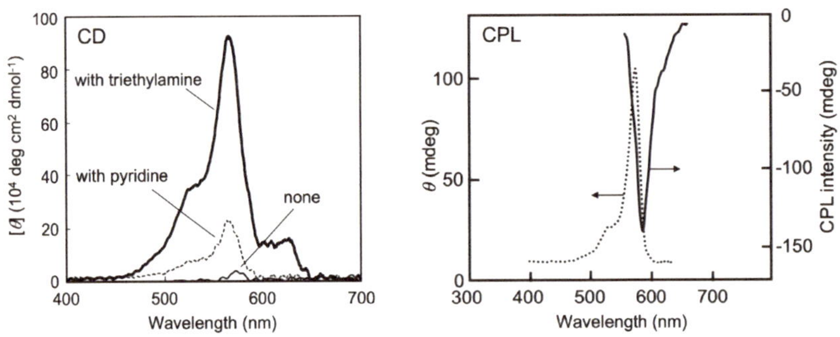

Figure 15.
Induction of CD and CPL in non-aqueous molecular gel system. [111] - reproduced by permission of the Royal Society of Chemistry.

its strength. Probably because of the insufficient acidity of the carboxyl group, neutralization of the anions in the cationic dye promoted a strong interaction between the dye and the chiral template.

3. Application for chiroptical polymer film

3.1 Encapsulation of chiral nanofiber into polymer

G-**Pyr** forms nanofibers in various polymerizable monomers, such as styrene, divinylbenzene, methylmethacrylate, and methylacrylate (**Figure 16**) [27]. For example, when *G*-**Pyr** is dissolved in methyl methacrylate at a concentration of 1 wt% and photopolymerized in a sandwich cell in the presence of a suitable photosensitizer, a transparent solid film can be obtained (**Figure 17**). When observing the CD spectrum of the obtained solid thin film, a CD signal showing a chiral orientation of pyrene groups was observed (**Figure 17**, red line). Since the strength and pattern are similar to those before polymerization (**Figure 17**, blue line), it is clear that the chiral orientation state of *G*-**Pyr** is fixed in the solid thin film. On the other hand, when photopolymerization was performed at 70 °C, a colorless and transparent solid thin film was also obtained, but almost no CD signal was detected (**Figure 17**, black line). That is, it is shown that the molecular orientation state cannot be obtained even if polymerization is carried out under the condition that the molecular orientation is not formed (nanofibers are not formed at 70 °C).

A more convenient method for producing a nanofiber composite polymer film is casting method that uses a polymer solution of general-purpose polymers such as polystyrene (PSt), polymethylmethacrylate (PMMA), and poly (ethylene-vinyl acetate) copolymers (EVA) [24, 82, 112–114]. **Figure 18** shows the CD and fluorescent spectra of a cast film prepared by spin coating from a PSt solution containing *G*-**Pyr**. **Figure 18** clearly shows that the excimer state (b) and the chiral stacking state (c) observed in the solution are reproduced in the polymer thin film [115].

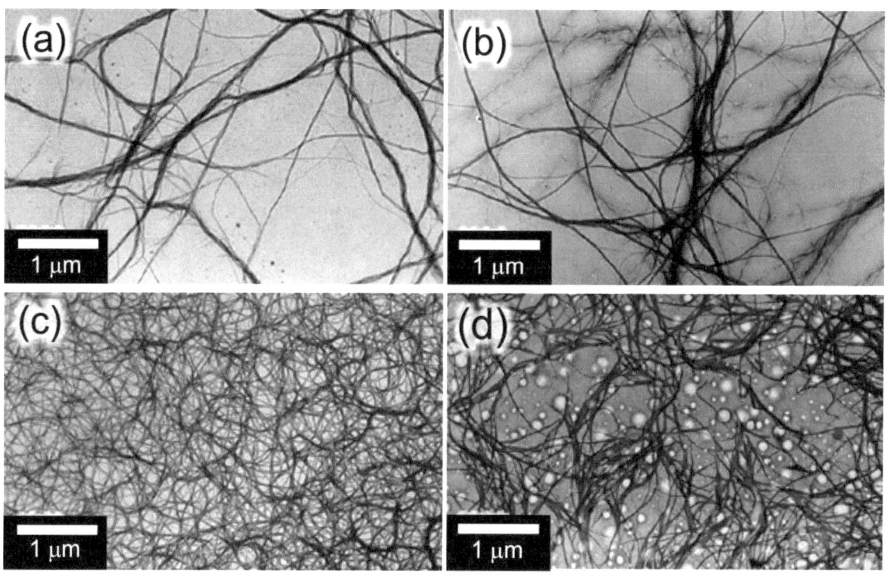

Figure 16.
*Aggregation morphology of G-**Pyr** in various polymerizable monomers: [27] (a) styrene; (b) divinylbenzene; (c) methylmethacrylate; and (d) methyl acrylate.*

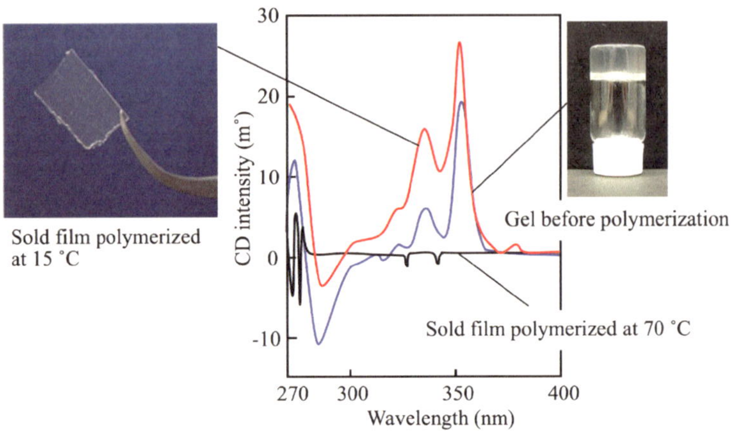

Figure 17.
*CD spectra of G-***Pyr** *in methylmethacrylate before and after polymerization [27].*

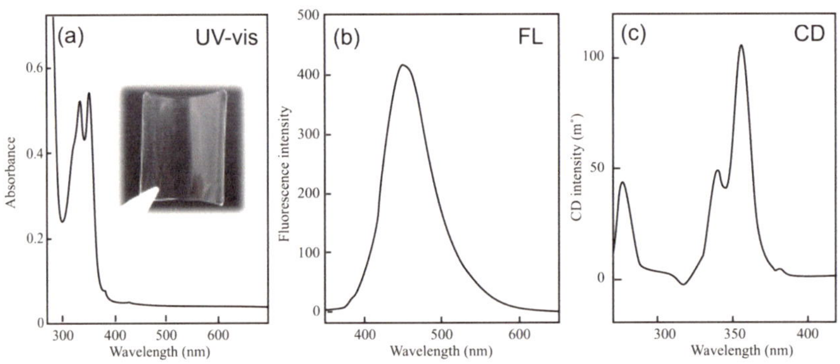

Figure 18.
*UV–visible (a), fluorescent (b) and CD (c) spectra of G-***Pyr***-containing polystyrene film prepared by casting from polymer solution. [115] - reproduced by permission of John Wiley and Sons.*

The formation of nanofibers derived from the molecular gelation phenomenon in the polymer can also be detected by direct observation with an electron microscope [112]. As shown in **Figure 19a**, when an EVA film containing *G*-**Pyr** is stained with an osmium plasma coater, it can be confirmed that nanofiber-like aggregates with a diameter of 10 nm or less are formed in the EVA film. On the other hand, for highly polar polymers, amphipathic *G*-$\textbf{Py}^{+}$ can be composited, and hollow nanofibers (**Figure 19b**) are formed in the polymer.

In the casting method, it is also possible to entrust the optical function to the added dye. **Figure 20** shows an example when an anionic molecular gel with *G*-**COOH** is used as the chiral nanofiber source. After preparing a transparent cast film consisting of a dye-nanofiber–polymer system to which three kinds of cationic dyes were added, distinct CD signals were detected around each absorption band of the added dyes (**Figure 20**). Because the dye does not have any chiral carbon or chirality by itself, the obtained CD signals cause the added dye to electrostatically bind to the chiral nanofibers in the polymer film. Therefore, they are induced CD [50].

3.2 Organic room temperature phosphorescent film

A transparent film, as shown in **Figure 21**, can be produced by preparing a mixed solution containing *G*-**BT** and EVA and forming a film by the casting

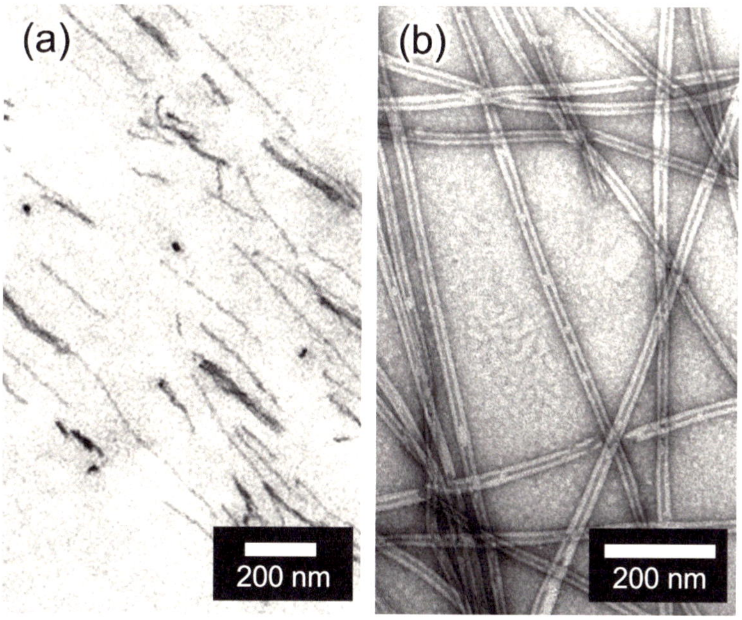

Figure 19.
*TEM images of (a) G-**Pyr** aggregates in poly(ethylene-vinylacetate) [112] and (b) G-**Py**$^+$ in polyvinylpyrrolidone [5]. Stained with (a) osmium and (b) uranyl acetate.*

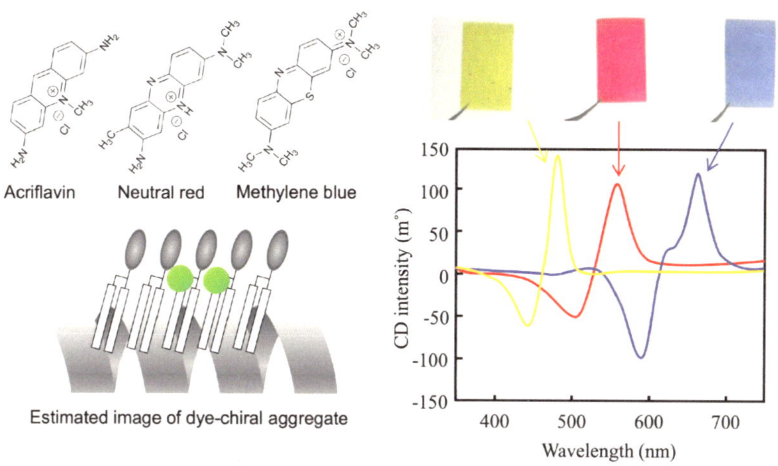

Figure 20.
Tuning of induced CD in dye-polymer mixed films [50].

method. The obtained film maintains a high Stokes shift and millisecond-order fluorescence lifetimes, likely to that of the solution systems. That is, the orientation state in the solution is also formed in the polymer. The formation of nanofibers in the polymer can be confirmed by laser microscopy in **Figure 21c** [83].

3.3 CPL polymer film

Figure 22a shows a glass plate in which polystyrene with a composite CPL source is cast on one side. The CPL source is a binary system using a molecular gel from *G*-**COOH** and a cationic cyanine dye [111]. Nanofibrils are detected using confocal microscopy in the polymer film, as shown in **Figure 22b**. **Figure 22c** shows

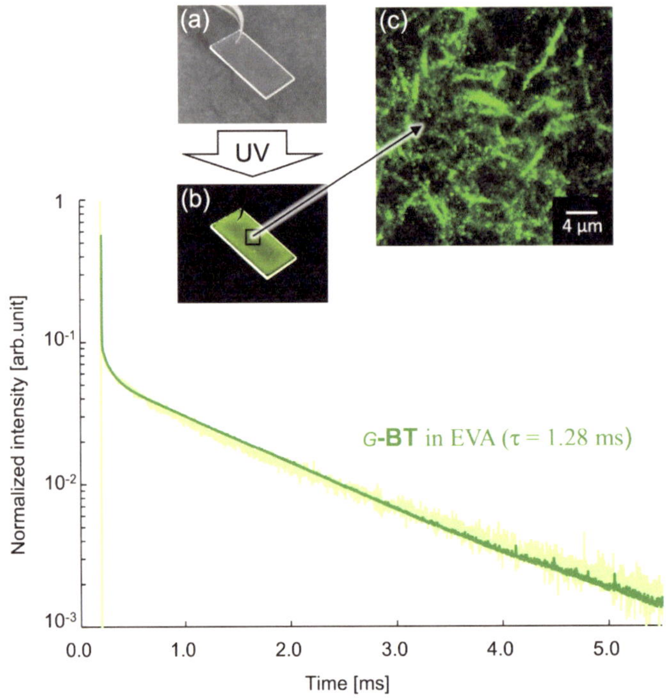

Figure 21.
*Example of room temperature phosphorescence by G-**BT**-containing polymer film. Luminescence decay curves were obtained at 525 nm. The photographs (a) and (b) are the G-BT-containing polymer film on quartz glass under normal and UV lights, respectively. (c) Confocal image of the G-**BT**-containing polymer film was obtained with excitation at 488 nm [83] - reproduced by permission of the Royal Society of Chemistry.*

the CPL spectrum of this film, in which the spectral shape and emission wavelength are similar to the results observed in its solution system. These facts indicate that the nanofiber and composite structure of the dye formed in the solution system were maintained as they were in the polymer.

Cyanine-based dyes, such as NK77 and 2012, are attractive as they combine with molecular gels to ensure good CPL strength; however, they are fragile due to light resistance. Therefore, binarization with a more chemically stable fluorescent dye is required. **Figure 23** shows the CPL spectra of a polymer film fabricated in combination with a more light-resistant dye, [111] ensuring good CPL strength and light resistance.

3.4 Application for wavelength conversion

The wavelength of sunlight and artificial lights cannot always be suitable for their applications, and light management using methods such as shading, wavelength cutting, and polarization is required depending on the application. In silicon-based solar cells, the spectral sensitivity to ultraviolet and near-infrared lights is low. Therefore, high energy levels of ultraviolet light are cut by glass. In order to effectively utilize unused light, it is necessary to introduce a technology that converts unnecessary ultraviolet and near-infrared light into visible light. Various fluorescent materials that use rare earths are leading the way in this field of application: nitride systems containing europium ions (Eu^{2+}) and cerium ions (Ce^{3+}) as activators, and rare earth materials such as garnet-based materials are used as fluorescent materials for lamps and white LEDs [116]. On the other hand, rare earth-free and low-toxic dyes are also required, and therefore, many organic fluorescent dyes have been

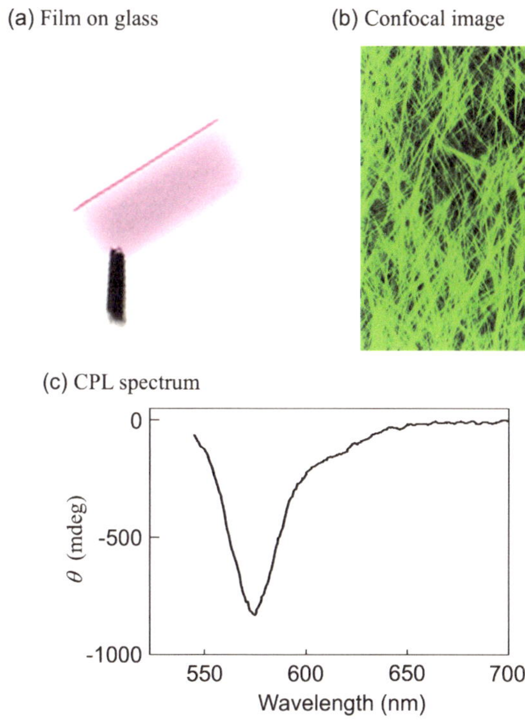

Figure 22.
*Aggregation morphology and CPL spectrum of the polystyrene composite film from the dye (NK-77) with G-**COOH** system. [111] - reproduced by permission of the Royal Society of Chemistry.*

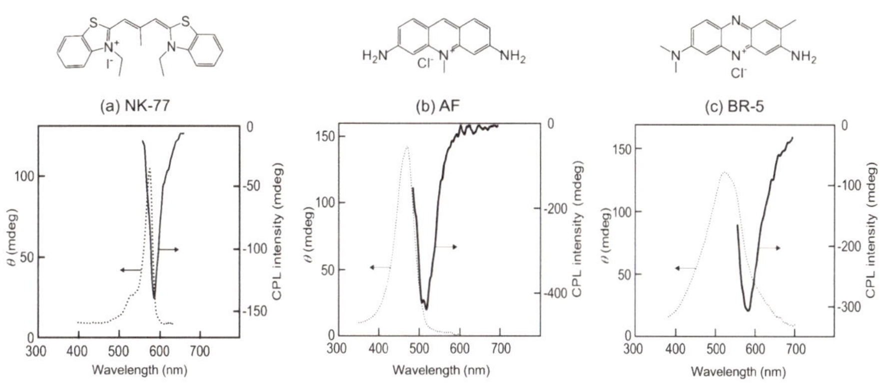

Figure 23.
Induction of strong CPL from the composite polymer film from various fluorescent dyes with chiral molecular gel system. [111] - reproduced by permission of the Royal Society of Chemistry.

developed [112, 115, 117–119]. Lightness, flexibility, and excellent processability are essential advantages of organic materials.

In this chapter, we will introduce the optical modulation function using self-assembled fluorescent nanofibers. As shown in **Figure** 7, a polymer film in which *G*-**Pyr** is embedded in polystyrene absorbing light in the UV-A region, which is not absorbed by ordinary inorganic glass, and it emits visible light. **Figure 24** shows an example in which a polymer system consisting of a *G*-**Pyr**-EVA composite is applied to the surface of a CIGS-based solar cell. Comparing the conversion efficiencies using simulated sunlight (AM1.5), it was confirmed that the power generation

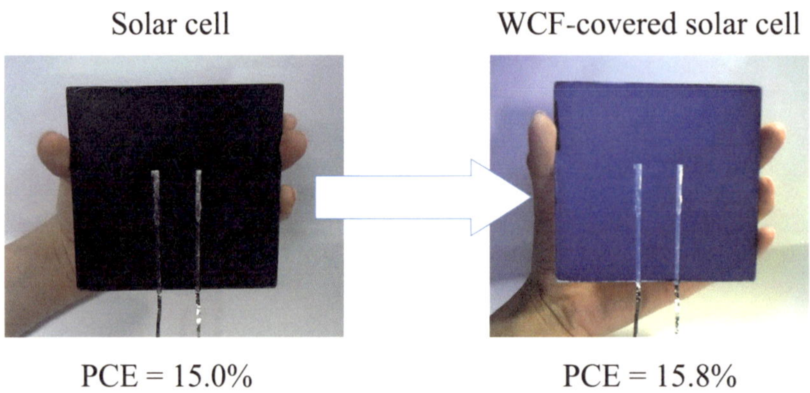

Figure 24.
Application of the wavelength conversion film (WCF) for CIGS solar cell. [6] - reproduced by permission of John Wiley and Sons.

efficiency before coating increased from 15.0% to 15.9% due to coating [112]. This increase in conversion efficiency is attributed to the fluorescent nanofibers embedded in the polymer film that absorb ultraviolet (UV-A region) light with low spectral sensitivity of the solar cell and have an emission peak in the visible region (~460 nm by excimer emission) with high spectral sensitivity.

4. Conclusions

In this chapter, we describe the characteristics of self-assembled nanofibers generated from amino acid-derived molecules, the expression principle of their unique optical properties, and their complexing with polymers. Since self-assembled nanofibers are structures formed by non-covalent bonds, one side that is physically fragile remains. However, conventional problems related to the dispersion of nanofibers are solved, and complicated dispersion techniques and surface modification processes that cause harmful effects are not required. Therefore, it is a material that enables higher-order functionalization of the polymer material while maintaining characteristics of the bulk polymer. In addition, various techniques for complementing vulnerabilities have been proposed. The optical management film with the optical modulation function introduced in this chapter is expected to be used not only for solar cells but also for various applications such as housing, automobiles, displays, artificial lighting, and plant factory lighting. We hope that the methodology using molecular gel-based functionalization can provide findings for further development.

Acknowledgements

This work was partially supported by a Grant-in-Aid for Scientific Research from the Ministry of Education, Culture, Sports, Science, and Technology, Japan.

Conflict of interest

The authors declare no conflict of interest.

DOI: http://dx.doi.org/10.5772/intechopen.96853

Author details

Hirotaka Ihara[1,2]*, Makoto Takafuji[2] and Yutaka Kuwahara[2]

1 National Institute of Technology, Okinawa College, Nago, Japan

2 Kumamoto University, Kumamoto, Japan

*Address all correspondence to: ihara@kumamoto-u.ac.jp

References

[1] Kargarzadeh H, Ahmad I, Thomas S, Dufresne A. Handbook of Nanocellulose and Cellulose Nanocomposites: Wiley-VCH; 2017. DOI: 10.1002/9783527689972

[2] Nakagaito AN, Yano H. The effect of morphological changes from pulp fiber towards nano-scale fibrillated cellulose on the mechanical properties of high-strength plant fiber based composites. Applied Physics A. 2004;78(4):547-552. DOI: 10.1007/s00339-003-2453-5

[3] Nakagaito AN, Yano H. Novel high-strength biocomposites based on microfibrillated cellulose having nano-order-unit web-like network structure. Applied Physics A. 2005;80(1):155-159. DOI: 10.1007/s00339-003-2225-2

[4] Ihara H, Takafuji M, Sakurai T. Self-assembled nanofibers. In: Nalwa HS, editor. Encyclopedia of Nanoscience & Nanotechnology. 9. California: American Scientific Publishers; 2004. p. 473-495.

[5] Ihara H, Takafuji M, Kuwahara Y. Polymer functionalization by luminescent supramolecular gels. Polymer Journal. 2016;48(8):843-853. DOI: 10.1038/pj.2016.53

[6] Ihara H, Takafuji M, Kuwahara Y, Okazaki Y, Ryu N, Sagawa T, Oda R. Supramolecular web and application for chiroptical functionalization of polymer. In: Yamamoto H, Kato T, editors. Molecular Technology. 4: WILEY-VCH; 2018. p. 297-337. DOI: 10.1002/9783527823987.vol4_c11

[7] Lin YC, Weiss RG. A novel gelator of organic liquids and the properties of its gels. Macromolecules. 1987;20(2):414-417. DOI: 10.1021/ma00168a031

[8] Ide N, Fukuda T, Miyamoto T. Gelation of fully acylated cellobiose in alkane solution. Bulletin of the Chemical Society of Japan. 1995;68(12):3423-3428. DOI: 10.1246/bcsj.68.3423

[9] Yamada K, Ihara H, Ide T, Fukumoto T, Hirayama C. Formation of helical super structure from single-walled bilayers by amphiphiles with oligo-L-glutamic acid-head group. Chemistry Letters. 1984;13(10):1713-1716. DOI: 10.1246/cl.1984.1713

[10] Ihara H, Hachisako H, Hirayama C, Yamada K. Lipid membrane analogues. Formation of highly-oriented structures and their phase separation behaviour in benzene. Journal of the Chemical Society, Chemical Communications. 1992(17):1244-1245. DOI: 10.1039/C39920001244

[11] Brotin T, Utermohlen R, Fages F, Bouaslaurent H, Desvergne JP. A novel small molecular luminescent gelling agent for alcohols. Journal of the Chemical Society, Chemical Communications. 1991; (6):416-418. DOI: 10.1039/c39910000416

[12] Ihara H, Takafuji M, Hirayama C, O'Brien DF. Effect of photopoly-merization on the morphology of helical supramolecular assemblies. Langmuir. 1992;8(6):1548-1553. DOI: 10.1021/la00042a010

[13] Hachisako H, Ihara H, Hirayama C, Yamada K. Chirally arranged monomeric dyes on helical bilayer membranes. Liquid Crystals. 1993;13(2):307-311. DOI: 10.1080/02678299308026304

[14] Arimura T, Shibata M, Ihara H, Hirayama C. Evaluation of selective binding ability in chiral supramolecules using induced chirality. Analytical Sciences. 1993;9(3):401-403. DOI: 10.2116/analsci.9.401

[15] Takafuji M, Ihara H, Hirayama C, Hachisako H, Yamada K. Functional organic gels - chirality induction through

formation of highly-oriented structure. Liquid Crystals. 1995;18(1):97-99. DOI: 10.1080/02678299508036596

[16] Ihara H, Shudo K, Hirayama C, Hachisako H, Yamada K. Functional organic gels. Enantioselective elution using chiral gels from amino acid-derived lipids. Liquid Crystals. 1996;20(6):807-809. DOI: 10.1080/02678299608033175

[17] Hachisako H, Ihara H, Yamada K. Supramolecular assemblies from amphiphilic oligomers with amino acid derived terminal groups. Recent Research Developments in Pure & Applied Chemistry. 2: Transworld Research Network; 1998. p. 59-81.

[18] Ihara H, Sakurai T, Yamada T, Hashimoto T, Takafuji M, Sagawa T, Hachisako H. Chirality control of self-assembling organogels from a lipophilic L-glutamide derivative with metal chlorides. Langmuir. 2002;18(19):7120-7123. DOI: 10.1021/la025535f

[19] Sagawa T, Chowdhury S, Takafuji M, Ihara H. Self-assembled nanofibrillar aggregates with amphiphilic and lipophilic molecules. Macromolecular Symposia. 2006;237:28-38. DOI: 10.1002/masy.200650504

[20] Gopal V, Prasad TK, Rao NM, Takafuji M, Rahman MM, Ihara H. Synthesis and in vitro evaluation of glutamide-containing cationic lipids for gene delivery. Bioconjugate Chemistry. 2006;17(6):1530-1536. DOI: 10.1021/bc0601020

[21] Kira Y, Okazaki Y, Sawada T, T akafuji M, Ihara H. Amphiphilic molecular gels from ω-aminoalkylated L-glutamic acid derivatives with unique chiroptical properties. Amino Acids. 2010;39(2):587-597. DOI: 10.1007/s00726-010-0480-z

[22] Jintoku H, Okazaki Y, Ono S, Takafuji M, Ihara H. Incorporation and template polymerization of styrene in single-walled bilayer membrane nanotubes. Chemistry Letters. 2011;40(6):561-563. DOI: 10.1246/cl.2011.561

[23] Okazaki Y, Jintoku H, Takafuji M, Oda R, Ihara H. Creation of a polymer backbone in lipid bilayer membrane-based nanotubes for morphological and microenvironmental stabilization. RSC Advances. 2014;4(63):33194-33197. DOI: 10.1039/c4ra03161j

[24] Goto T, Okazaki Y, Ueki M, Kuwahara Y, Takafuji M, Oda R, Ihara H. Induction of strong and tunable circularly polarized luminescence of nonchiral, nonmetal, low-molecular-weight fluorophores using chiral nanotemplates. Angewandte Chemie International Edition. 2017;56(11):2989-2993. DOI: 10.1002/anie.201612331

[25] Ihara H, Yoshitake M, Takafuji M, Yamada T, Sagawa T, Hirayama C, Hachisako H. Detection of highly oriented aggregation of L-glutamic acid-derived lipids in dilute organic solution. Liquid Crystals. 1999;26(7):1021-1027. DOI: 10.1080/026782999204363

[26] Sagawa T, Fukugawa S, Yamada T, Ihara H. Self-assembled fibrillar networks through highly oriented aggregates of porphyrin and pyrene substituted by dialkyl L-glutamine in organic media. Langmuir. 2002;18(19):7223-7228. DOI: 10.1021/la0255267

[27] Takafuji M, Ishiodori A, Yamada T, Sakurai T, Ihara H. Stabilization of enhanced chirality from pyrene-containing L-glutamide lipid in methyl methacrylate by photo-induced polymerization. Chemical Communications. 2004;10(9):1122-1123. DOI: 10.1039/b316673b

[28] Ihara H, Yamada T, Nishihara M, Sakurai T, Takafuji M, Hachisako H, Sagawa T. Reversible gelation in cyclohexane of pyrene substituted by dialkyl L-glutamide: Photophysics of

the self-assembled fibrillar network. Journal of Molecular Liquids. 2004;111(1-3):73-76. DOI: 10.1016/S0167-7322(03)00263-0

[29] Hachisako H, Ihara H, Kamiya T, Hirayama C, Yamada K. Thermal isomerization process in benzene gels of L-glutamic acid-derived lipids with spiropyran head groups. Chemical Communications. 1997;(1):19-20. DOI: 10.1039/a606386a

[30] Hachisako H, Nakayama H, Ihara H. Determination of critical aggregation concentrations of self-assembling lipids in nonpolar organic media using spiropyrans as photochromic probes. Chemistry Letters. 1999;28(11):1165-1166. DOI: 10.1246/cl.1999.1165

[31] Takafuji M, Shirosaki T, Yamada T, Sakurai T, Alekperov D, Popova G, Sagawa T, Ihara H. Dendritic cyclotriphosphazene derivative with hexaxis(alkylazobenzene) substitution as photo-sensitive trigger. Heterocycles. 2004;63(7):1563-1572. DOI: 10.3987/com-04-10077

[32] Yamada T, Derakhshan M, Ansarian HR, Takafuji M, Hachisako H, Sagawa T, Ihara H. Self-assembly-based thermo-responsible luminescent organogels of chromophoric L-glutamide-derived lipids. Journal of Materials Research. 2005;20(9):2486-2490. DOI: 10.1557/JMR.2005.0296

[33] Shirosaki T, Chowdhury S, Takafuji M, Alekperov D, Popova G, Hachisako H, Ihara H. Functional organogels from lipophilic L-glutamide derivative immobolized on cyclotriphosphazene core. Journal of Materials Research. 2006;21(5):1274-1278. DOI: 10.1557/jmr.2006.0156

[34] Hachisako H, Ryu N, Hashimoto H, Murakami R. Formation of specific dipolar microenvironments complementary to dipolar betaine dye by nonionic peptide lipids in nonpolar medium. Organic & Biomolecular Chemistry. 2009;7(11):2338-2346. DOI: 10.1039/B818218C

[35] Takafuji M, Kira Y, Ishiordori A, Hachisako H, Sawada T, Ihara H. Enclosure of secondary chirality based on highly-oriented lipid aggregates into a polymer sheet by photo-induced polymerization of polymerizable monomer gels. Macromolecular Symposia. 2010;291-292(1):330-336. DOI: 10.1002/masy.201050539

[36] Miyamoto K, Jintoku H, Sawada T, Takafuji M, Sagawa T, Ihara H. Informative secondary chiroptics in binary molecular organogel systems for donor-acceptor energy transfer. Tetrahedron Letters. 2011;52(31):4030-4035. DOI: 10.1016/j.tetlet.2011.05.131

[37] Wang X, Duan P, Liu M. Self-assembly of π-conjugated gelators into emissive chiral nanotubes: Emission enhancement and chiral detection. Chemistry – An Asian Journal. 2014;9(3):770-778. DOI: 10.1002/asia.201301518

[38] Duan P, Zhu X, Liu M. Isomeric effect in the self-assembly of pyridine-containing L-glutamic lipid: Substituent position controlled morphology and supramolecular chirality. Chemical Communications. 2011;47(19):5569-5571. DOI: 10.1039/C1CC10813A

[39] Nakashima N, Yamaguchi Y, Eda H, Kunitake M, Manabe O. Design of a lipid bilayer electrical device. Strong chemical structure dependence and molecular mechanisms on the phase transition-dependent electrical impedance responses of the device in air. The Journal of Physical Chemistry B. 1997;101(2):215-220. DOI: 10.1021/jp963221t

[40] Miyamoto K, Sawada T, Jintoku H, Takafuji M, Sagawa T, Ihara H. Controlled emission enhancement and quenching by self-assembly of low molecular weight thiophene derivatives.

Tetrahedron Letters. 2010;51(35):4666-4669. DOI: 10.1016/j.tetlet.2010.07.006

[41] Ihara H, Takafuji M, Sakurai T, Katsumoto M, Ushijima N, Shirosaki T, Hachisako H. Novel self-assembling organogelators by combination of a double chain-alkylated L-glutamide and a polymeric head group. Organic and Biomolecular Chemistry. 2003;1(17): 3004-3006. DOI: 10.1039/b305928f

[42] Zhou X, Jin Q, Zhang L, Shen Z, Jiang L, Liu M. Self-assembly of hierarchical chiral nanostructures based on metal–benzimidazole interactions: Chiral nanofibers, nanotubes, and microtubular flowers. Small. 2016;12(34):4743-4752. DOI: 10.1002/smll.201600842

[43] Zhang L, Jin Q, Liu M. Enantio-selective recognition by chiral supramolecular gels. Chemistry – An Asian Journal. 2016;11(19):2642-2649. DOI: 10.1002/asia.201600441

[44] Zhang L, Wang X, Wang T, Liu M. Tuning soft nanostructures in self-assembled supramolecular gels: From morphology control to morphology-dependent functions. Small. 2015;11(9-10):1025-1038. DOI: 10.1002/smll.201402075

[45] Takafuji M, Sakurai T, Yamada T, Hashimoto T, Kido N, Sagawa T, Hachisako H, Ihara H. Metal ion-induced chirality and morphology control of self-assembling organogels from L-glutamic acid-derived lipids. Chemistry Letters. 2002;31(5):548-549. DOI: 10.1246/cl.2002.548

[46] Miao W, Zhang L, Wang X, Cao H, Jin Q, Liu M. A dual-functional metallogel of amphiphilic copper(II) quinolinol: Redox responsiveness and enantioselectivity. Chemistry – A European Journal. 2013;19(9):3029-3036. DOI: 10.1002/chem.201203401

[47] Hachisako H, Murakami R. Intense fluorescence-inducing amphiphile in cationic dyes and its applicability. Chemical Communications. 2006;(10):1073-1075. DOI: 10.1039/B517483J

[48] Hachisako H, Murata Y, Ihara H. Supramolecular receptors from α-amino acid-derived lipids. Journal of the Chemical Society, Perkin Transactions 2. 1999; (11):2569-2577. DOI: 10.1039/a903956b

[49] Hatano T, Bae AH, Takeuchi M, Fujita N, Kaneko K, Ihara H, Takafuji M, Shinkai S. Helical structures of conjugate polymers created by oxidative polymerization using synthetic lipid assemblies as templates. Chemistry - A European Journal. 2004;10(20):5067-5075. DOI: 10.1002/chem.200400332

[50] Takafuji M, Kira Y, Tsuji H, Sawada S, Hachisako H, Ihara H. Optically active polymer film tuned by a chirally self-assembled molecular organogel. Tetrahedron. 2007;63(31):7489-7494. DOI: 10.1016/j.tet.2007.02.036

[51] Hachisako H, Motozato Y, Murakami R, Yamada K. Extraordinary monomer-dimer transition of methylene blue induced by the phase transition of telomer-bilayer membranes formed from dialkyl L-glutamate amphiphiles with oligo-acrylic acid-head group. Chemistry Letters. 1992;21(2):219-222. DOI: 10.1246/cl.1992.219

[52] Takafuji M, Azuma N, Miyamoto K, Maeda S, Ihara H. Polycondensation and stabilization of chirally ordered molecular organogels derived from alkoxysilyl group- Containing L-glutamide lipid. Langmuir. 2009;25(15):8428-8433. DOI: 10.1021/la804321u

[53] Hachisako H, Ryu N, Murakami R. Molecular structural requirements, dye specificity, and application of anionic peptide amphiphiles that induce intense fluorescence in cationic dyes. Organic & Biomolecular Chemistry. 2009;7(11):2327-2337. DOI: 10.1039/B818206J

[54] Ihara H, Hachisako H, Hirayama C, Yamada K. Specific bindings of methyl orange to chiral bilayer membranes with β-alanyl-L-glutamoyl head groups [1]. Liquid Crystals. 1987;2(2):215-221. DOI: 10.1080/02678298708086292

[55] Watanabe N, Jintoku H, Sagawa T, Takafuji M, Sawada T, Ihara H. Self-assembling fullerene derivatives for energy transfer in molecular gel system. Journal of Physics: Conference Series. 2009;159. DOI: 10.1088/1742-6596/159/1/012016

[56] Ryu N, Hachisako H. Functionalization of methyl orange using cationic peptide amphiphile: colorimetric discrimination between ATP and ADP at pH 2.0. Organic & Biomolecular Chemistry. 2011;9(6):2000-2006. DOI: 10.1039/C0OB00437E

[57] Jintoku H, Shimoda S, Takafuji M, Sagawa T, Ihara H. Tuning of molecular orientation of porphyrin assembly according to monitoring the chiroptical signals. Molecular Crystals and Liquid Crystals. 2011;539:403-407. DOI: 10.1080/15421406.2011.566057

[58] Hatano T, Bae AH, Takeuchi M, Fujita N, Kaneko K, Ihara H, Takafuji M, Shinkai S. Helical superstructure of conductive polymers as created by electrochemical polymerization by using synthetic lipid assemblies as a template. Angewandte Chemie - International Edition. 2004;43(4):465-469. DOI: 10.1002/anie.200351749

[59] Ihara H, Ide T, Hotta O, Uchio H, Hirayama C, Yamada K. Amphiphiles with polypeptide-head groups III. Regulation of enantioselectivity in micellar hydrolysis. Polymer Journal. 1986;18(6):463-469. DOI: 10.1295/polymj.18.463

[60] Ihara H, Fukumoto T, Hirayama C, Yamada K. Exceptional morphologies and metamorphosis of bilayer membranes formed from amphiphiles with poly(L-aspartic acid)-head groups. Polymer communications. 1986;27(9):282-285. DOI:

[61] Hirayama C, Ihara H. Development of macroporous polymer packings for aqueous high flow rate liquid chromatography. Approach on the viewpoint of materials chemistry. Analytical Sciences. 1991;7(4):527-536. DOI: 10.2116/analsci.7.527

[62] Gopal V, Xavier J, Kamal MZ, Govindarajan S, Takafuji M, Soga S, Ueno T, Ihara H, Rao NM. Synthesis and transfection efficiency of cationic oligopeptide lipids: Role of linker. Bioconjugate Chemistry. 2011;22(11):2244-2254. DOI: 10.1021/bc2002874

[63] Duan P, Li Y, Li L, Deng J, Liu M. Multiresponsive chiroptical switch of an azobenzene-containing lipid: Solvent, temperature, and photoregulated supramolecular chirality. The Journal of Physical Chemistry B. 2011;115(13):3322-3329. DOI: 10.1021/jp110636b

[64] Xue P, Sun J, Xu Q, Lu R, Takafuji M, Ihara H. Anion response of organogels: Dependence on intermolecular interactions between gelators. Organic & Biomolecular Chemistry. 2013;11(11):1840-1847. DOI: 10.1039/C3OB27241A

[65] Xue P, Lu R, Zhang P, Jia J, Xu Q, Zhang T, Takafuji M, Ihara H. Amplifying emission enhancement and proton response in a two-component gel. Langmuir. 2013;29(1):417-425. DOI: 10.1021/la3037617

[66] Jintoku H, Kao M-T, Del Guerzo A, Yoshigashima Y, Masunaga T, Takafuji M, Ihara H. Tunable Stokes shift and circularly polarized luminescence by supramolecular gel. Journal of Materials Chemistry C.

2015;3(23):5970-5975. DOI: 10.1039/C5TC00878F

[67] Takafuji M, Kawahara T, Sultana N, Ryu N, Yoshida K, Kuwahara Y, Oda R, Ihara H. Extreme enhancement of secondary chirality through coordination-driven steric changes of terpyridyl ligand in glutamide-based molecular gels. RSC Advances. 2020;10(50):29627-29632. DOI: 10.1039/D0RA05057A

[68] Goldstein AS, Amory JK, Martin SM, Vernon C, Matsumoto A, Yager P. Testosterone delivery using glutamide-based complex high axial ratio microstructures. Bioorganic & Medicinal Chemistry. 2001;9(11):2819-2825. DOI: 10.1016/S0968-0896(01)00149-3

[69] Jintoku H, Sagawa T, Sawada T, Takafuji M, Ihara H. Versatile chiroptics of peptide-induced assemblies of metalloporphyrins. Organic and Biomolecular Chemistry. 2010;8(6):1344-1350. DOI: 10.1039/b920058d

[70] Jintoku H, Sagawa T, Takafuji M, Ihara H. Noncovalent one-to-one donor-acceptor assembled systems based on porphyrin molecular gels for unusually high electron-transfer efficiency. Chemistry - A European Journal. 2011;17(41):11628-11636. DOI: 10.1002/chem.201101043

[71] Jintoku H, Takafuji M, Oda R, Ihara H. Enantioselective recognition by a highly ordered porphyrin-assembly on a chiral molecular gel. Chemical Communications. 2012;48(40):4881-4883. DOI: 10.1039/c2cc31127e

[72] Jintoku H, Sagawa T, Miyamoto K, Takafuji M, Ihara H. Highly efficient and switchable electron-transfer system realised by peptide-assisted J-type assembly of porphyrin. Chemical Communications. 2010;46(38):7208-7210. DOI: 10.1039/c0cc01190h

[73] Lee J-h, Jintoku H, Okazaki Y, Sagawa T, Takafuji M, Ihara H. Manipulation of discrete porphyrin–fullerene nanopillar arrays regulated by the phase separated infiltration of polymer in ternary blended organic thin-films. Solar Energy Materials and Solar Cells. 2015;140:428-438. DOI: 10.1016/j.solmat.2015.04.043

[74] Yang X, Zhang G, Zhang D, Xiang J, Yang G, Zhu D. Self-assembly of a new C60 compound with a L-glutamid-derived lipid unit: formation of organogels and hierarchically structured spherical particles. Soft Matter. 2011;7(7):3592-3598. DOI: 10.1039/C0SM01109F

[75] Yang X, Zhang G, Zhang D, Zhu D. A new ex-TTF-based organogelator: Formation of organogels and tuning with fullerene. Langmuir. 2010;26(14):11720-11725. DOI: 10.1021/la101193z

[76] Babu SS, Prasanthkumar S, Ajayaghosh A. Self-assembled gelators for organic electronics. Angewandte Chemie International Edition. 2012;51(8):1766-1776. DOI: 10.1002/anie.201106767

[77] Kunitake T, Ihara H, Okahata Y. Phase separation and reactivity changes of phenyl ester substrate and imidazole catalyst in the dialkylammonium bilayer membrane. Journal of the American Chemical Society. 1983;105(19):6070-6078. DOI: 10.1021/ja00357a018

[78] Terech P, Weiss RG. Low molecular mass gelators of organic liquids and the properties of their Gels. Chemical Reviews. 1997;97(8):3133-3160. DOI: 10.1021/cr9700282

[79] van Esch JH, Feringa BL. New functional materials based on self-assembling organogels: from serendipity towards design. Angewandte Chemie International Edition. 2000;39(13):2263-2266. DOI: 10.1002/1521-3773(20000703)39:13<2263::AID-ANIE2263>3.0.CO;2-V

[80] Shimizu T, Masuda M, Minamikawa H. Supramolecular nanotube architectures based on amphiphilic molecules. Chemical Reviews. 2005;105(4):1401-1444. DOI: 10.1021/cr030072j

[81] Okazaki Y, Goto T, Sakaguchi R, Kuwahara Y, Takafuji M, Oda R, Ihara H. Facile and versatile approach for generating circularly polarized luminescence by non-chiral, low-molecular dye-on-nanotemplate composite system. Chemistry Letters. 2016;45(4):448-450. DOI: 10.1246/cl.160047

[82] Ihara H, Yamaguchi M, Takafuji M, Hachisako H, Hirayama C, Yamada K. Production of helical bilayer membranes from L-glutamic acid derivatives with bis(dodecylamide) groups and their specific optical activity. Journal of the Chemical Society of Japan. 1990;1990(10):1047-1053. DOI: 10.1246/nikkashi.1990.1047

[83] Yoshida K, Kuwahara Y, Miyamoto K, Nakashima S, Jintoku H, Takafuji M, Ihara H. A room-temperature phosphorescent polymer film containing a molecular web based on one-dimensional chiral stacking of a simple luminophore. Chemical Communications. 2017;53(36):5044-5047. DOI: 10.1039/c7cc00395a

[84] Shibata M, Ihara H, Hirayama C. Unique property of cyanine dyes on charged poly(L-lysine). Polymer. 1993;34(5):1106-1108. DOI: 10.1016/0032-3861(93)90238-6

[85] Sagawa T, Tobata H, Ihara H. Exciton interactions in cyanine dye-hyaluronic acid (HA) complex: Reversible and biphasic molecular switching of chromophores induced by random coil-to-double-helix phase transition of HA. Chemical Communications. 2004;10(18):2090-2091. DOI: 10.1039/b408032g

[86] Okazaki Y, Ryu N, Buffeteau T, Pathan S, Nagaoka S, Pouget E, Nlate S, Ihara H, Oda R. Induced circular dichroism of monoatomic anions: Silica-assisted the transfer of chiral environment from molecular assembled nanohelices to halide ions. Chemical Communications. 2018;54(73):10244-10247. DOI: 10.1039/c8cc05449e

[87] Mashima S, Ryu N, Kuwahara Y, Takafuji M, Jintoku H, Oda R, Ihara H. Multi-chiro-informative system created by a porphyrin-functionalized chiral molecular assembly. Chemistry Letters. 2020;49(4):368-371. DOI: 10.1246/cl.200018

[88] Haraguchi S, Numata M, Li C, Nakano Y, Fujiki M, Shinkai S. Circularly polarized luminescence from supramolecular chiral complexes of achiral conjugated polymers and a neutral polysaccharide. Chemistry Letters. 2009;38(3):254-255. DOI: 10.1246/cl.2009.254

[89] Tsumatori H, Nakashima T, Kawai T. Observation of chiral aggregate growth of perylene derivative in opaque solution by circularly polarized luminescence. Organic Letters. 2010;12(10):2362-2365. DOI: 10.1021/ol100701w

[90] Kaseyama T, Furumi S, Zhang X, Tanaka K, Takeuchi M. Hierarchical assembly of a phthalhydrazide-functionalized helicene. Angewandte Chemie International Edition. 2011;50(16):3684-3687. DOI: 10.1002/anie.201007849

[91] Langhals H, Hofer A, Bernhard S, Siegel JS, Mayer P. Axially chiral bichromophoric fluorescent dyes. The Journal of Organic Chemistry. 2011;76(3):990-992. DOI: 10.1021/jo102254a

[92] Sawada Y, Furumi S, Takai A, Takeuchi M, Noguchi K, Tanaka K. Rhodium-catalyzed enantioselective synthesis, crystal structures, and

photophysical properties of helically chiral 1,1′-bitriphenylenes. Journal of the American Chemical Society. 2012;134(9):4080-4083. DOI: 10.1021/ja300278e

[93] Ikeda T, Masuda T, Hirao T, Yuasa J, Tsumatori H, Kawai T, Haino T. Circular dichroism and circularly polarized luminescence triggered by self-assembly of tris(phenylisoxazolyl) benzenes possessing a perylenebisimide moiety. Chemical Communications. 2012;48(48):6025-6027. DOI: 10.1039/C2CC31512B

[94] Kumar J, Nakashima T, Tsumatori H, Mori M, Naito M, Kawai T. Circularly polarized luminescence in supramolecular assemblies of chiral bichromophoric perylene bisimides. Chemistry – A European Journal. 2013;19(42):14090-14097. DOI: 10.1002/chem.201302146

[95] Morisaki Y, Gon M, Sasamori T, Tokitoh N, Chujo Y. Planar chiral tetrasubstituted [2.2]paracyclophane: Optical resolution and functionalization. Journal of the American Chemical Society. 2014;136(9):3350-3353. DOI: 10.1021/ja412197j

[96] Nakamura K, Furumi S, Takeuchi M, Shibuya T, Tanaka K. Enantioselective synthesis and enhanced circularly polarized luminescence of S-shaped double azahelicenes. Journal of the American Chemical Society. 2014; 136(15):5555-5558. DOI: 10.1021/ja500841f

[97] Shiraki T, Tsuchiya Y, Noguchi T, Tamaru S, Suzuki N, Taguchi M, Fujiki M, Shinkai S. Creation of circularly polarized luminescence from an achiral polyfluorene derivative through complexation with helix-forming polysaccharides: Importance of the meta-linkage chain for helix formation. Chemistry – An Asian Journal. 2014;9(1):218-222. DOI: 10.1002/asia.201301216

[98] Inouye M, Hayashi K, Yonenaga Y, Itou T, Fujimoto K, Uchida T-a, Iwamura M, Nozaki K. A doubly alkynylpyrene-threaded [4]rotaxane that exhibits strong circularly polarized luminescence from the spatially restricted excimer. Angewandte Chemie International Edition. 2014;53(52):14392-14396. DOI: 10.1002/anie.201408193

[99] Gon M, Morisaki Y, Chujo Y. Optically active cyclic compounds based on planar chiral [2.2]paracyclophane: Extension of the conjugated systems and chiroptical properties. Journal of Materials Chemistry C. 2015;3(3):521-529. DOI: 10.1039/C4TC02339K

[100] Shen Z, Wang T, Shi L, Tang Z, Liu M. Strong circularly polarized luminescence from the supramolecular gels of an achiral gelator: Tunable intensity and handedness. Chemical Science. 2015;6(7):4267-4272. DOI: 10.1039/C5SC01056J

[101] Rybicka A, Longhi G, Castiglioni E, Abbate S, Dzwolak W, Babenko V, Pecul M. Thioflavin T: Electronic circular dichroism and circularly polarized luminescence induced by amyloid fibrils. Chemphyschem. 2016;17(18):2931-2937. DOI: 10.1002/cphc.201600235

[102] Ye Q, Zhu D, Xu L, Lu X, Lu Q. The fabrication of helical fibers with circularly polarized luminescence via ionic linkage of binaphthol and tetraphenylethylene derivatives. Journal of Materials Chemistry C. 2016;4(7):1497-1503. DOI: 10.1039/C5TC04174K

[103] Liu S, Li F, Wang Y, Li X, Zhu C, Cheng Y. Circularly polarized luminescence of chiral 1,8-naphthalimide-based pyrene fluorophore induced via supramolecular self-assembly. Journal of Materials Chemistry C. 2017;5(24):6030-6036. DOI: 10.1039/C7TC01371J

[104] Huang G, Wen R, Wang Z, Li BS, Tang BZ. Novel chiral aggregation

induced emission molecules: Self-assembly, circularly polarized luminescence and copper(II) ion detection. Materials Chemistry Frontiers. 2018; 2(10):1884-1892. DOI: 10.1039/c8qm00294k

[105] Ma Z, Winands T, Liang N, Meng D, Jiang W, Doltsinis NL, Wang Z. A C2-symmetric triple [5]helicene based on N-annulated triperylene hexaimide for chiroptical electronics. Science China Chemistry. 2019;63(2):208-214. DOI: 10.1007/s11426-019-9632-2

[106] Miki K, Noda T, Gon M, Tanaka K, Chujo Y, Mizuhata Y, Tokitoh N, Ohe K. Near-infrared circularly polarized luminescence through intramolecular excimer formation of oligo(p-phenyleneethynylene)-based double helicates. Chemistry – A European Journal. 2019;25(39):9211-9216. DOI: 10.1002/chem.201901467

[107] Zhang S, Fan J, Wang Y, Li D, Jia X, Yuan Y, Cheng Y. Tunable aggregation-induced circularly polarized luminescence of chiral AIEgens via the regulation of mono–/di-substituents of molecules or nanostructures of self-assemblies. Materials Chemistry Frontiers. 2019;3(10):2066-2071. DOI: 10.1039/c9qm00358d

[108] Dhbaibi K, Favereau L, Srebro-Hooper M, Quinton C, Vanthuyne N, Arrico L, Roisnel T, Jamoussi B, Poriel C, Cabanetos C, Autschbach J, Crassous J. Modulation of circularly polarized luminescence through excited-state symmetry breaking and interbranched exciton coupling in helical push-pull organic systems. Chemical Science. 2020;11(2):567-576. DOI: 10.1039/c9sc05231c

[109] Frédéric L, Desmarchelier A, Plais R, Lavnevich L, Muller G, Villafuerte C, Clavier G, Quesnel E, RacineB,Meunier-Della-GattaS,DognonJP, ThuéryP,CrassousJ,FavereauL,PietersG. Maximizing chiral perturbation on thermally activated delayed fluorescence emitters and elaboration of the first top-emission circularly polarized OLED. Advanced Functional Materials. 2020;30(43). DOI: 10.1002/adfm.202004838

[110] Takaishi K, Iwachido K, Ema T. Solvent-induced sign inversion of circularly polarized luminescence: Control of excimer chirality by hydrogen bonding. Journal of the American Chemical Society. 2020;142(4):1774-1779. DOI: 10.1021/jacs.9b13184

[111] Oishi H, Mashima S, Kuwahara Y, Takafuji M, Yoshida K, Oda R, Qiu H, Ihara H. Polymer encapsulation and stabilization of molecular gel-based chiroptical information for strong, tunable circularly polarized luminescence film. Journal of Materials Chemistry C. 2020;8(26):8732-8735. DOI: 10.1039/d0tc01480j

[112] Jintoku H, Yamaguchi M, Takafuji M, Ihara H. Molecular gelation-induced functional phase separation in polymer film for energy transfer spectral conversion. Advanced Functional Materials. 2014;24(26):4105-4112. DOI: 10.1002/adfm.201304081

[113] Jintoku H, Dateki M, Takafuji M, Ihara H. Supramolecular gel-functionalized polymer films with tunable optical activity. Journal of Materials Chemistry C. 2015;3(7):1480-1483. DOI: 10.1039/c4tc02948h

[114] Oishi H, Yoshida K, Kuwahara Y, Takafuji M, Oda R, Ihara H. Generation of strong circularly polarized luminescence induced by chiral organogel based on L-glutamide. Journal of the Taiwan Institute of Chemical Engineers. 2018;92:58-62. DOI: 10.1016/j.jtice.2018.03.026

[115] Jintoku H, Ihara H. Molecular gel-mediated UV-to-visible spectral conversion for enhancement of

power-conversion efficiency. Chemical Communications. 2012;48(8):1144-1146. DOI: 10.1039/c2cc16911h

[116] Justel T, Nikol H, Ronda C. New developments in the field of luminescent materials for lighting and displays. Angewandte Chemie International Edition. 1998;37(22):3084-3103. DOI: 10.1002/(SICI)1521-3773(19981204)37:22<3084::AID-ANIE3084>3.0.CO;2-W

[117] Wakita J, Inoue S, Kawanishi N, Ando S. Excited-state intramolecular proton transfer in imide compounds and its application to control the emission colors of highly fluorescent polyimides. Macromolecules. 2010;43(8):3594-3605. DOI: 10.1021/ma100126n

[118] Park S, Kim S, Seo J, Park SY. Strongly fluorescent and thermally stable functional polybenzoxazole film: Excited-state intramolecular proton transfer and chemically amplified photopatterning. Macromolecules. 2005;38(11):4557-4559. DOI: 10.1021/ma050009r

[119] Chu Q, Medvetz DA, Pang Y. A polymeric colorimetric sensor with excited-state intramolecular proton transfer for anionic species. Chemistry of Materials. 2007;19(26):6421-6429. DOI: 10.1021/cm0713982

Chapter 8

Chirality in Anticancer Agents

Jindra Valentová and Lucia Lintnerová

Abstract

Many drugs are chiral and their therapeutic activity depends on specific recognition of chiral biomolecules. The biological activity of enantiomers can also differ drastically in terms of toxicity and pharmacokinetics. Chiral natural biological molecules, such as nucleic acids, enzymes are targeted molecules for the development of anticancer drugs. The interest in chiral agents is logically a result of the different interaction with biomolecules leading in the end consequence to improve anticancer activity and maybe to less undesirable effects. This review outlines the effects of chirality on the efficiency of anticancer metal-based agents and potential organic drugs. A variety of up-to-date examples of structurally diverse chiral agents exhibiting different mechanisms in their antitumor activity is presented.

Keywords: Chirality, anticancer, metal based, chiral complexes, drugs, organic metal agents

1. Introduction

Regardless of the origin, chirality is an integral part of biological process that derived their inherent asymmetry from the chirality of the fundamental building blocks of receptors – the L-amino acids. It would thus be expected that a receptor protein derived from the enantiomeric D-amino acids would have the same fundamental properties, but exhibit opposing chirality of interaction. In addition, the macromoleculecular structures of biopolymers also give rise to chirality as a results of helicity. Such helical structures may have either a left or righ handed turn. In the case of DNA double helix and the protein α-helix, the biopolymers have a right-handed turn. As nature has a preference in terms of its chirality, enzymes and receptor system exhibit stereochemical preferences, particularly as many of natural substrates and ligands for these system, e.g. neurotrasmitters, endogenous opioids, and hormones [1].

It is now not surprising, that stereoisomers of drugs may differ in their pharmacodynamic activity at their biological target, or in their pharmacokinetic properties (absorption, distribution, and clearance by metabolism and excretion) [2, 3]. Enantiomers may differ both quantitatively and qualitatively in their biological activities. At one extreme, one enantiomer may be devoid of any biological activity, at the other extreme, both enantiomers may have qualitatively different biological activities. These stereoselective differences may arise not only from drug interactions at the pharmacological receptors, but also from pharmacokinetic events [4, 5].

Advances in chemical technology, especially in the methodology of stereoselective syntheses and stereospecific analyzes, together with regulatory measures, have led to an increase in the number of new authorized chirally pure drugs over the last ten years, and this trend persists [6]. In contrast to the end of the last century, when about 55% of clinically used drugs were chiral and half of them were used as racemates, the current trend in the development of new chiral drugs is mainly towards substances containing pure one enantiomeric form [7, 8]. Currently, chiral formations of small molecules are an essential part of the discovery and development of new anticancer medicines [2, 9].

2. Anticancer therapy

Cancer is a large group of diseases that can start in almost any organ or tissue of the body when abnormal cells grow uncontrollably and have the potential to invade or spread to other parts of the body. Cancer cells are formed when normal cells lose the normal regulatory mechanism that controls their growth and multiplication. If the cancer cells remain localized they are said to be *benign.* If the cancer cells invade other part of the body and set up secondary tumors a process known as metastasis - the cancer is defined as *malignant* and is life threatening [10]. A major problem in treating cancer is the fact that it is not a single disease. There are more than 200 different cancers resulting from different cellular effects, and so a treatment that is effective in controling one type of cancer may be ineffective on another [11].

Cancer is the second leading cause of death globally, behind only ischemic heart disease. Lung, prostate, colorectal, stomach and liver cancer are the most common types of cancer in men, while breast, colorectal, lung, cervical and thyroid cancer are the most common among women [12].

The primary treatment modalities include surgery, chemotherapy, radiation, and immunotherapy, etc. However, mainstay treatment is based on chemotherapy which is a viable alternative involving various natural and synthetic compounds that can kill or stop the unwanted proliferation of cancerous cells [13, 14].

The compounds used in the chemotherapy of cancer diseases are quite varied in structure and mechanism of action, including alkylating agents, antimetabolite analogs of folic acid, pyrimidine and purine; natural products, hormones and hormone antagonist; and a variety of agents directed at specific molecular targets. Most these of agents interacted with DNA or its precursors, inhibiting the synthesis of new genetic material and causing damage to the DNA of both normal and malignant cells. Rapidly expanding knowledge about cancer biology has led to the discovery of entirely new and more cancer specific targets (e.g. growth factor receptors, intracelluar signaling pathways, epigenic processes, tumor vascularity, DNA repairs effect, and cell death pathway [15].

3. Chiral metal-based anticancer agents

Metal-based antitumor chemotherapeutics obtained prominence after the discovery of the cytotoxic effect of cisplatin - cis-diamminedichloroplatinum (II), *cis*-[Pt $(NH_3)_2Cl_2$] – **Figure 1**. Currently, cisplatin is one of the most effective chemotherapeutic drug used for many solid malignancies [16]. The discovery of cisplatin stimulated efforts to investigate other platinum and non platinum metal-containing compounds for their potential use in the treatment of cancer. In recent years, more attention has been devoted to complexes of other transition metals (ruthenium, gold, osmium, iridium, etc.) which offer a different mechanism of action and lower systemic toxicity compared to platinum-based drugs [17, 18].

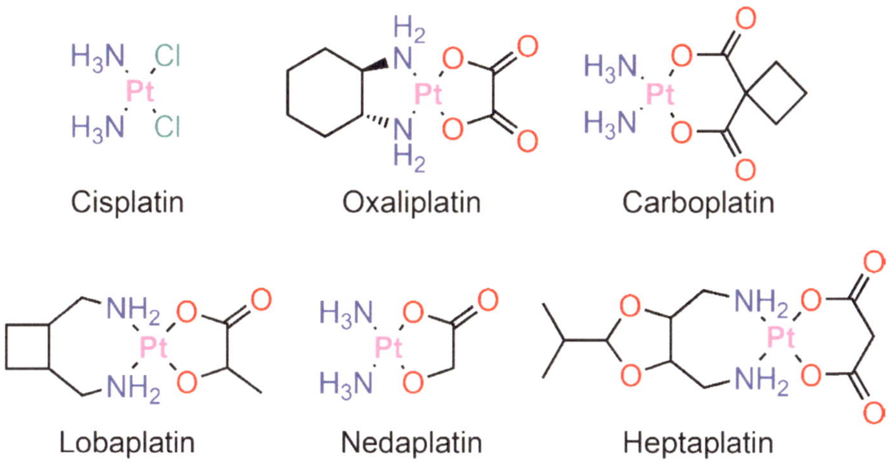

Figure 1.
The chemical structures of clinically used platinum-based anticancer drugs.

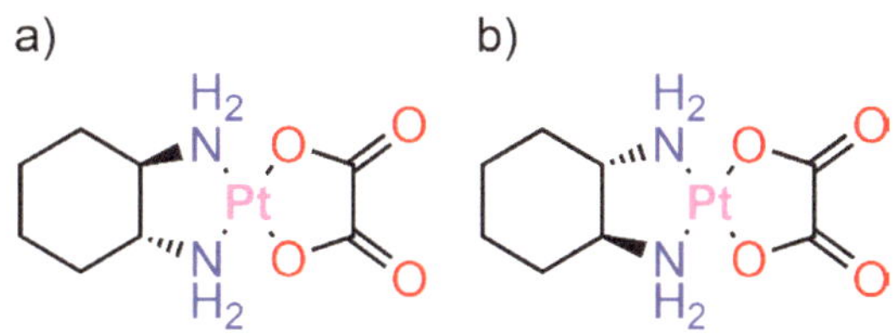

Figure 2.
The structures oxaliplatin with chiral ligands (a) 1R,2R-*cyclohexane-1,2-diamine; (b)* 1S,2S-*cyclohexane-1,2-diamine.*

Chirality is one of the important paradigms that decide the precise structure of a particular anticancer drug and its interaction with the cancer molecular target. This is valid not only for organic drugs but also for metal-based drugs which are developed for anticancer application [19, 20]. An, example is oxaliplatin, which contains the chiral ligand (*R, R*)-cyclohexane-1,2-diamine. This platinum antineoplastic compound is more biologically active in comparison with its enantiomer contaning the ligand (*S, S*)-cyclohexane-1,2-diamine **Figure 2** [20]. Interest in the development of chiral pure metal-based compounds for anticancer application is increasing.

Chiral metal-based anticancer drugs have been comprehensively reviewed in the literature [9, 21] therefore we will focus on selected examples to give the reader an overview of the most important developments. This review is limited to a variety of some recent examples of structurally diverse chiral agents exhibiting different mechanism in their anticancer effect.

Chirality in metal coordination complexes can be observed in two ways:

1. The metal centre has an asymmetric arrangement of ligands around it. This type of chirality can be observed predominantly in octahedral complexes and tetrahedral complexes,

2. The metal centre has a chiral ligand (i.e. the ligand itself has a non-superimposable mirror image).

Various methods have been used to denote the absolute configuration of optical isomers such as R or S, Λ or Δ, or C and A [22]. According to IUPAC rules the

R/S convention is applied to enantiomers with tetrahedral geometry. In the case of compounds with geometries other than tetrahedral, the same principles are followed as for the C/A convention (C for the clockwise and A for the anticlockwise priority sequence of ligands), and for bis and tris bidentate complexes the absolute configuration is designated Lambda Λ (left-handed) and Delta Δ (right-handed). Conformers with helical chirality are designated as δ (right-handed helix) and λ (left-handed helix).

3.1 Chiral Pt-based anticancer complexes

Since cisplatin was firstly approved by the US Food nd Drug Administration (FDA) in 1978, platinum complexes have become a class of important anticancer drugs [23]. Cisplatin has been widely employed to treat a variety of tumors including ovarian, cervical, head and neck, non-small cell lung carcinoma, and testicular cancers, and is commonly used in combination regimens [24]. However, the clinical application of platinum drugs has several drawback including serious toxicity and, natural and acquired drug resistance [25, 26]. Therefore, much effort has been made to overcome the side effects and improve the antitumor activity of platinum-based drugs [27–31].

To date, second and third generation platinum analogues, namely carboplatin and oxaliplatin, have been designated and approved for worldwide use. Carboplatin is effective in the treatment of ovarian carcinoma, lung, and head and neck cancers, while oxaliplatin is clinically approved for the treatment of colorectal cancer, which is resistant to cisplatin [32].

Other platinum compounds have regional approval, nedaplatin is used in Japan, lobaplatin is applied for treatment of chronic myeloid leukemia (China), heptaplatin is used for treatment of stomach cancer (in Korea). [32], though there is a little information about influence of its stereochemistry on biological effect **Figure 1** [33].

It is widely accepted that the main biochemical mechanism of action of platinum (II) drugs involves the binding of the drug to DNA in the cell nucleus and subsequent interference with normal transcription, and/or DNA replication mechanisms [34]. Cis platin becomes activated once it enters the cell. In the cytoplasm, chloride atoms on cisplatin are displaced by water molecules. This hydrolysed product is a potent electrophile that can react with any nucleophile, including the sulfhydryl groups on proteins and nitrogen donor atoms on nucleic acids. Cisplatin binds to the N7 reactive centre on purine residues and as such can cause deoxyribonucleic acid (DNA) damage in cancer cells, blocking cell division and resulting in apoptotic cell death [35]. Some studies, using spectroscopics techniques and crystallography, have shown an energetically favourable interaction between the chiral ligands of platinum complexes stereoisomers and the DNA structure or small nucleotides [36–38].

Platinum drugs derived from chiral 1,2-diaminecyclohexane ligand (DACH) – were the first class of compounds used to investigate the relations between spatial configuration of the ligand and the ability of the complexes to form efficient adducts with DNA [39, 40].

Each DACH stereoisomer has a particularly stable bulk structure with a nearly planar shape for (*RR*) and (*SS*)-(DACH) dichloroplatinum (II) and an L-shaped molecule for the (*RS*) stereoisomer. As a consequence of these conformations, when the RS stereoisomer forms an adduct with DNA, significant steric hindrance is expected, resulting in a slower binding of the complexes to DNA as well as a different recognition and processing of the adducts formed. This theory has been confirmed by numerous cytotoxicity tests (predominantly towards human tumor cell lines) on DACH and their structural analogues which have, in most cases, shown the same order of activity (RR > SS > RS) [41]. From the series of platinum drugs with DACH ligands only [$PtCl_2$ (*R, R*-DACH) has been approved for clinical use [25].

Sometimes, when small modifications are made in the structure of chemically related compounds (for instance: a change in the substitution pattern of an aromatic ring or an alkyl chain) or in the conditions used to test the complexes (for instance: a change in the cell lines, a change from an *in vitro* to an *in vivo* test), a reversal of the stereoselectivity order can be detected [42, 43]. These findings suggest the possibility that interaction of platinum complexes with other biomolecules might be responsible for their stereoselectivity. These compounds are in most instances amino acids, proteins present in biological fluids and taking part in the distribution and metabolism of platinum complexes [41]. A difference in cellular accumulation between diastereoisomers was found in the platinum complex with phenyl derivative of chiral diamine ligands **Figure 3** [44]. These findings supported the hypothesis that stereoselectivity in cellular transport can also cause a different accumulation of platinum complexes in kidney and a different toxic effect of isomers [45].

The direct targeting of the G-quadruplex structure of DNA which is widely present in human telomeric DNA, transcription start sites, and promoter regions of genes responsible for cell apoptosis/growth and senescence, might play a key role in the anticancer action of chiral metal(II) complexes [46, 47]. The G-quadruplex structure is important in the control of a variety of cellular processes, including telomere maintenance, gene replication, transcription and translation [48, 49]. The induction/stabilization of quadruplex DNA is considered to be a potential target for the development of anticancer drugs [50, 51].

Platinum(II) complexes with chiral 4-(2,3-dihydroxypropyl)-formamide oxoaporphine R/S-(±)-FOA (1), R-(+)-FOA (2) and S-(−)-FOA ligands – **Figure 4**, showed significant antitumor activity caused by directly targeting G-quadruplex DNA [52]. Among these platinum(II) complexes, the complex with S-(−)-FOA was found to exhibit higher selectivity and telomerase inhibition via targeting telomere G4s in BEL-7404 cells, as well as inducing S phase arrest and telomeres/DNA damage, which resulted in cell senescence and apoptosis. Similar results were also found in ruthenium(II) complexes with these chiral FOA ligands. The *in vitro* anticancer activity of the Ru(II) complex with *S*-(−)-FOA was higher compare to complex with -(±)-FOA (1), *R*-(+)-FOA [53].

Apart from platinum compounds, a number of other anticancer metal complexes are currently being investigated. The main advantage of complexes containing a metal other than platinum is the different mechanism of action compared

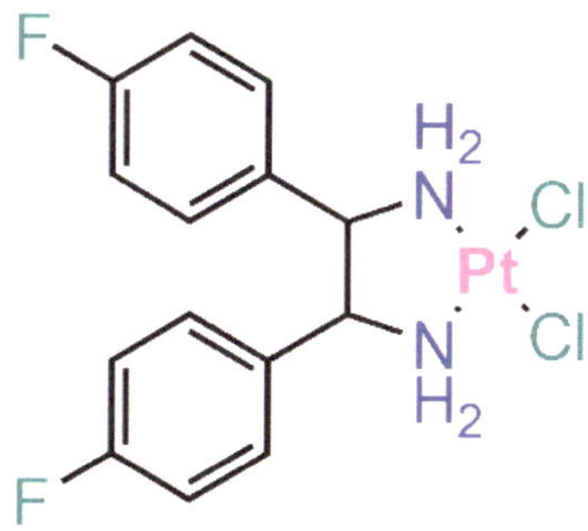

Figure 3.
Examples of Platinum(II) complexes with chiral diamines ligands where stereochemical discrimination of cytotoxic activity was SS > RR > meso.

to platinum-based chemotherapy, as well as a wider range of treated tumors, the ability to overcome resistance, and lower systemic toxicity [54, 55].

3.2 Chiral ruthenium anticancer complexes

Ruthenium complexes are promising candicates as chemotherapeutic agents because some of them have exhibited favorable *in vitro* and *in vivo* pharmacological profiles in different models including platinum-resistant cells [56, 57]. For instance, the ligand exchange kinetics of Pt(II) and Ru(II) complexes in aqueous solution, which is crucial for anticancer activity, are very similar; thus Ru is considered to be an alternative to platinum-based drugs. Many ruthenium compounds are less toxic than Pt-based drugs and some of them are quite selective for cancer cells. These facts may arise from the ability of ruthenium to mimic iron in binding to biomolecules [56, 58]. A number of ruthenium(II/III)-based anticancer agents have been developed to date, yet none of them are in clinical use as anticancer drugs. Currently, two Ru(III) complexes are in clinical studies: NAMI-A and KP1339 type ruthenium(III) coordination complexes based on N, Cl donor ligands **Figure 5** [59].

The influence of chirality on anticancer effect has been reported primarily in the group of polypyridyl and organometallic Ru complexes and [9]. Wang et al. [60] presented a series chiral polypyridyl complexes (Δ-Ru1, Λ-Ru1, Δ-Ru2, Λ-Ru2, Δ-Ru3, and Λ-Ru3) – **Figure 6** as mitochondria targeting anticancer agents. The cytotoxic activity of these ruthenium(II) complexes was tested against six selected

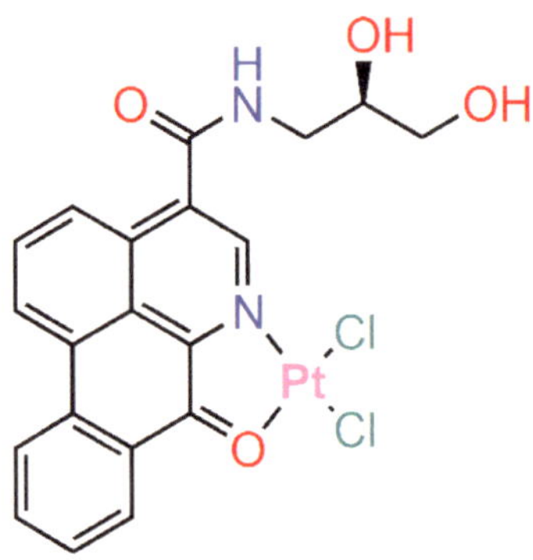

Figure 4.
The structures of chiral Platinum(II) complexes with chiral FOA ligand as G4-DNA and telomerase inhibitor.

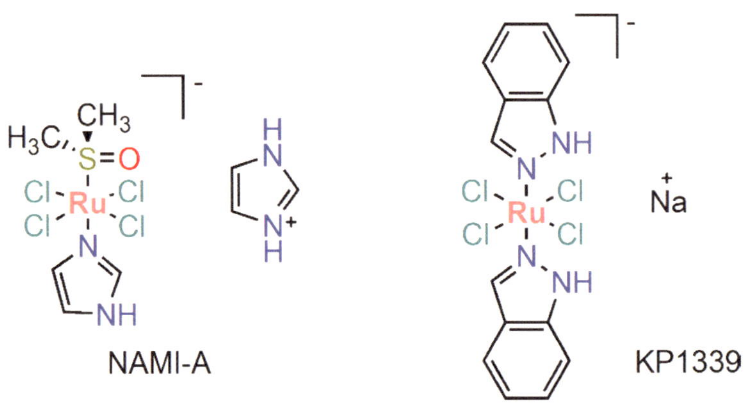

Figure 5.
The structures of ruthenium complexes in clinical studies.

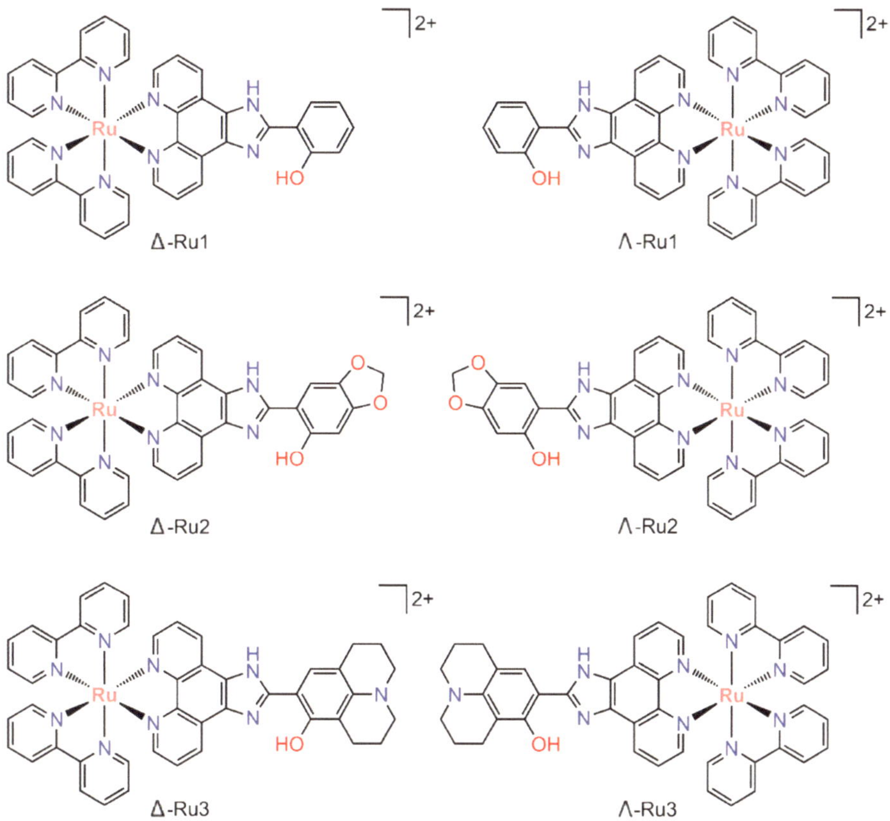

Figure 6.
The structures of chiral polypyridyl Ru(II) complexes.

human cancer cell lines, (HeLa, A549, HepG2, MCF-7, BEL-7402, and MG-63), and one cisplatin resistant cell line. Δ-Ru1 was the most active anticancer agent, exhibiting cytotoxicity similar to that of cisplatin against the six cancer cell lines. Notably, this complex inhibited the growth of the cisplatin-resistant cell line A549-CP/R, suggesting that the anticancer mechanism of Δ-Ru1 differs from that of cisplatin. Λ-Ru1 displayed slightly lower toxicity than Δ-Ru1, which also seems to be attributed to the slightly less cellular uptake of ruthenium. In HeLa cells, the same relationship was found for Δ-Ru2 and Λ-Ru2, but the opposite relationship was found for Δ-Ru3 and Λ-Ru3.

Further analysis showed that Δ-Ru1 exerts its toxicity through the intrinsic mitochondria-mediated apoptotic pathway, which is accompanied by the regulation of Bcl-2 family members and the activation of caspases [60].

Some studies have highlighted the potential of polypyridyl Ru complexes to target G4 and telomerase [61, 62]. Sun et al. [63] revealed that enatinomers of Ruthenium complexes $[Ru(phen)2(p\text{-}DMNP)]^{2+}$ – **Figure 7** possessed binding affinities and significant selectivity for human G-quadruplex DNA. The Λ–Ru complext can stabilize human telomeric G quadruplex DNA slightly more than Δ–Ru and has a strong preference for G-quadruplex over duplex DNA. The specific recognition of HepG2 cells makes Λ–Ru complex a promising class of luminescent labels for (bio) imaging agents.

Chiral organometallic ruthenium (II) complexes have been studied as protein kinase inhibitors [64, 65]. In the case of chiral complexes containing bidentate staurosporine ligand – **Figure 8** the (*R*)-enantiomer was a significantly more potent inhibitor in comparison to the (*S*)-enantiomer [65].

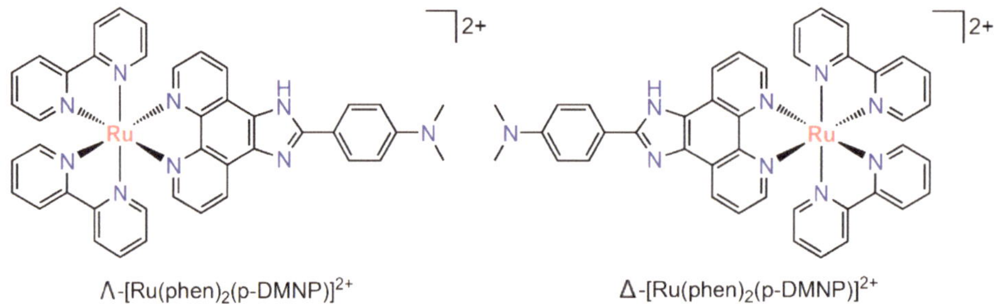

Figure 7.
Examples of chiral polypyridyl Ru(II) complexes targeting G4 and telomerase.

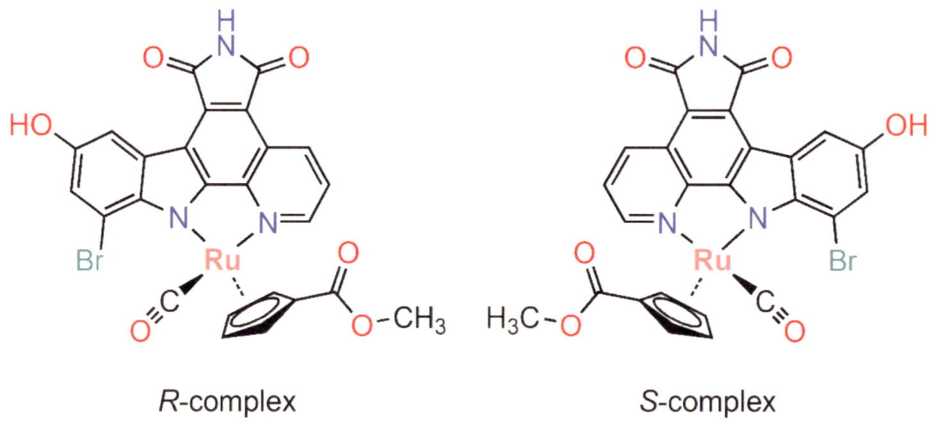

Figure 8.
The chemical structures of chiral Ru complexes with enantiomers of R- and S-staurosporine ligand.

Half-sandwich organometallic arene complexes have also been studied for anticancer application, though studies of pure stereoisomers of arene complexes are limited, probably due to difficulties in their resolution [21]. Sadler et al. described the cytotoxicity of pure diastereoisomers of Ru arene complexes - **Figure 9** [9, 66]. The *R,S* or *S,R* isomers showed higher cytotoxicity against the human ovarian cancer cell line (A2780) than the *R,R* or *S,S* compounds. However, the mechanism of the different activities between the enantiomers was not reported.

Cuvea-Alique and co-workers [67] evaluated the anticancer activities of arene Ru(II) enantiomers witzzh aminooxime ligands- **Figure 10**. The oxime-containing Ru(II) compounds have shown potent anticancer activities against a broad range of different cancer cell lines, with no significant differences between the two enantiomers.

Other chiral organometallic complexes that have been studied for their anticancer properties include osmium [68, 69], iridium [70, 71], rhodium [72] and gold [73] complexes.

3.3 Chiral gold anticancer complexes

Gold complexes have emerged as a versatile and effective class of metal-based anticancer agents [74, 75]. Auranofin, (2,3,4,6- tetra-*O*-acetyl-1-(thio-*κ*S)-β-L-glucopyranosato)- (triethylphosphine) gold(I), is an orally administrated anti-arthritis drug which has also been clinically studied for its antiproliferative properties **Figure 11** [74].

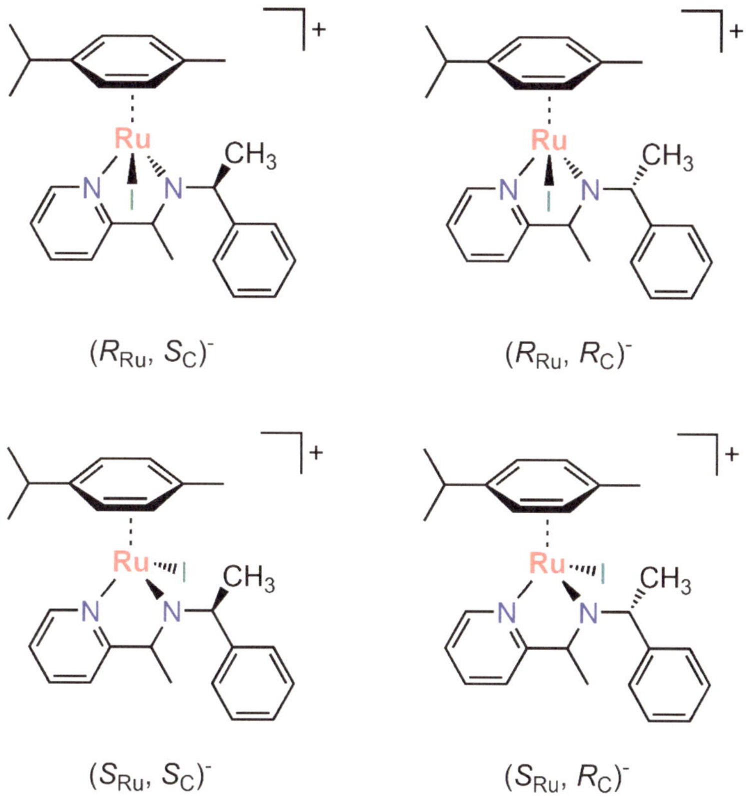

Figure 9.
The structures of diastereoisomers of organometallic arene Ru (II) complexes.

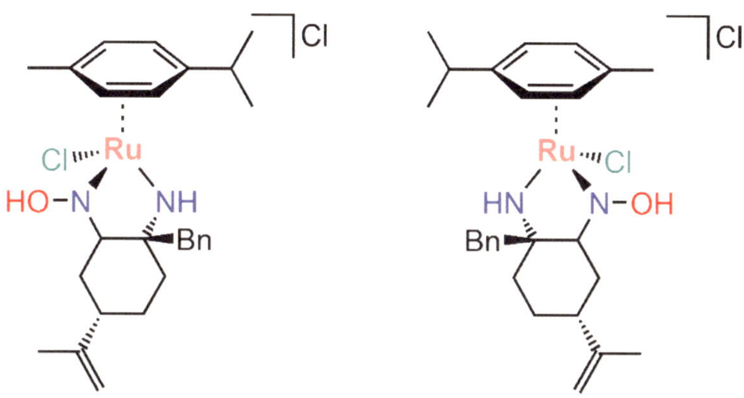

Figure 10.
The structures of chiral organometallic arene Ru (II) complexes with aminoxime ligand.

Based on these results different Au(I) complexes bearing stereogenic phosphine [76, 77], chiral N-heterocyclic carbenes ligands (NHC) [78, 79] have been synthesized and tested as potential anticancer drugs.

Enantiomeric Au(I) complexes with a phospohorus stereogenic centre - **Figure 12** has shown great toxicity against both suspended and adherent cancer cells but with marked difference s in their toxicity to healthy cells. The (*R*,*R*)-enantiomers were more toxic against healthy mammalian cells compare to those of the (*S*,*S*)-enantiomer [76, 77].

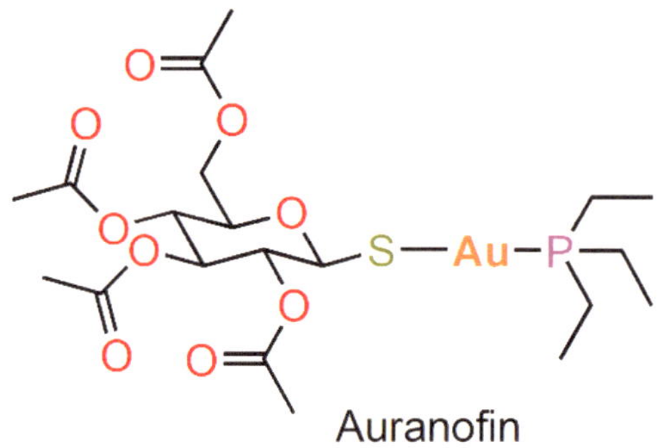

Figure 11.
The structure of gold complex, which is under clinical investigation for treatment cancer.

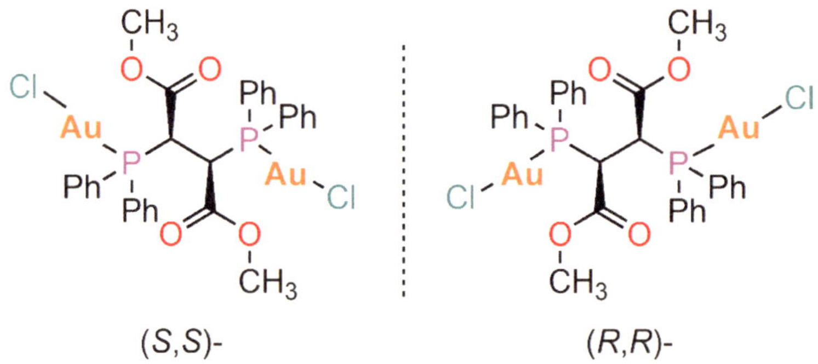

Figure 12.
Examples of enantiomeric phosphine-Au(I) complexes.

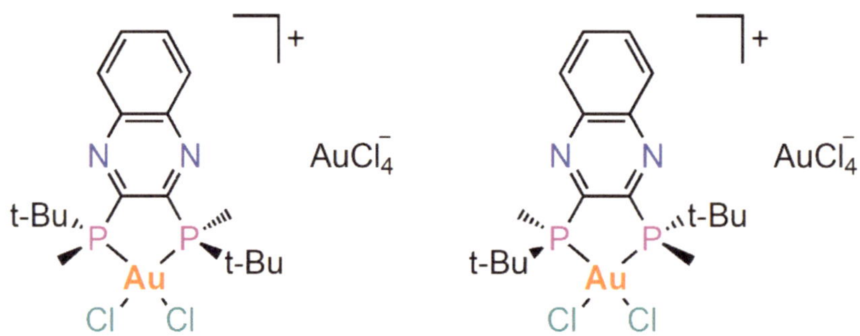

Figure 13.
Chemical structures of the enantiomeric phosphine-Au(III) complexes.

The mechanism of action of Au(I) complexes is thought to trigger apoptosis *via* inhibition of selenium- and sulfur-containing enzymes such as thioredoxin reductase (TrxR), glutathione peroxidase, cysteine protease or glutathione-S-transferase [80].

The novel chiral Au(III) complexes containing chiral P stereogenic phosphine ligand: *R,R* and *S,S*- QuinoxP* (2,3-bis(tert-butylmethylphoshino)quinoxaline) were prepared **Figure 13** [81]. The antiproliferative activities of the two compound were evaluated against panel of cell lines and exhibited cytotoxicity slightly higher than that of cisplatin and auranofin. However, no diffrence was found in the cytotoxic effect between the two enantiomers.

4. Chiral organic anticancer agents

The search for novel natural and synthetic compounds with potential antitumor activity is a major goal of many research groups. The tests involve known compounds with proven biological activity that have not been previously used in oncology as well as newly synthesized molecules that are structural analogues of biologically active substances already used in the treatment of cancer [82].

4.1 Alkylating agents

Alkylating agents were among the first group of chemicals determined to be useful in cancer chemotherapy, and the largest drug group among conventional cytotoxic chemotherapeutics. They are so named because of their ability to add alkyl groups to negatively charged groups on biological molecules such as DNA and proteins [83]. Classical alkylating agents include nitrogen mustards, nitroso ureas, aziridines, and alkyl sulfonates. Nonclassical alkylating agents include hydrazine, triazene, and altretamines. In addition, the alkylating-like agent group, which similarly to alkylating agents functions by crosslinking with DNA, includes platinum compounds. The high clinical efficacy of one of the main classes of alkylating agents, nitrogen mustards, makes them the current choice for the first-line treatment of different tumor types. However, severe side effects limit the therapeutic value of such compounds, and new effective compounds are required [84].

The group of bifunctional chloropiperidine derivatives has been revealed to possess novel efficient DNA-alkylating properties, leading to direct strand cleavage at guanine nucleotides and indirect effects on the human topoisomerase II enzyme [85, 86].

4.1.1 Chiral chloropiperidines

Carraro et al. [87] investigated series of racemic and enantiomerically pure monofunctional chloropiperidines – **Figure 14**. Derivatives of chloropiperidines demonstrated the ability to alkylate DNA *in vitro*. On a panel of carcinoma cell lines, M-CePs exhibited low nanomolar cytotoxicity indexes, which showed their remarkable activity against pancreatic cancer cells and in all cases performed strikingly better than the chlorambucil control. Interestingly, stereochemistry modulated the activity of the chloropiperidines.

An analysis of the cytotoxicity of enantiomers *N*-butyl derivatives of chloropiperidines D-1 and L-1 in three cancer cell lines revealed that D-1 was the most

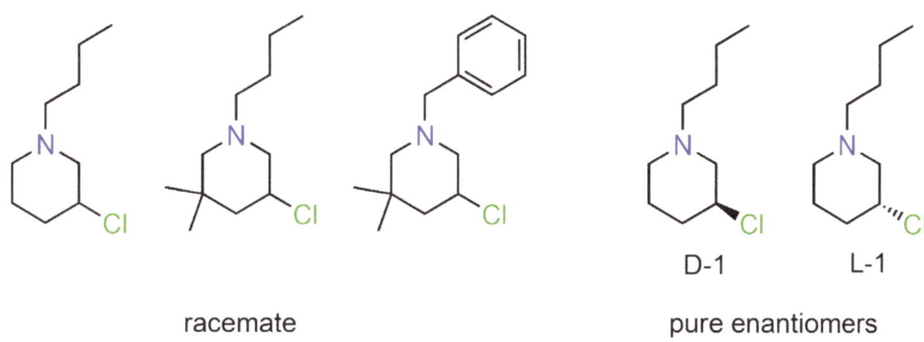

Figure 14.
Chemical structure of racemic chloropiperidines and enatiomerically pure compounds.

active compound, again with a clear tropism for pancreatic cancer cells, while its enantiomer, L-1 enantiomer was less cytotoxic, with an eudismic ratio of ~40 in the case of BxPC-3 cells. The direct damage observed to isolated DNA was not sufficient to explain their nanomolar cytotoxicity, especially when considering the enantiomeric couple.

The D-1 enantiomer turned out to be the most cytotoxic compound of the entire series, although it was inactive in the DNA cleavage assay. In contrast, its mirror image, L-1, found to efficiently nick and fragment the plasmid *in vitro*, happened to be much less cytotoxic.

The enantiomers also differed in permeation through an artificial membrane, that simulates passive diffusion. Because D-1 was the most cytotoxic but least permeable enantiomer, the permeation analysis of the chiral compounds suggests the involvement of active mechanisms of uptake into cells.

4.2 Inhibitors of topoisomerases

4.2.1 Chiral epoxy-substituted chromones

Human DNA topoisomerases (Top) have been recognized as a good target molecule for the development of anticancer drugs because they play an important role in solving DNA topological problems caused by DNA strand separation during replication and transcription [88].

Jo et al. [89] designed and synthesized novel chiral epoxy-substituted chromone analogues -**Figure 15** that exhibit an anticancer effect by inhibiting the DNA synthesis of cancer cells. Their ability to alkylate DNA and inhibit topo enzymes and cancer cell growth was evaluated.

In the brief structure–activity relationship analysis, no clear correlation was seen between stereochemistry and topos inhibitory and cytotoxic activity. However, compounds **6**, **10** and **11** were more potent than the others in both Top I and IIα inhibitory activity.

The 5(*R*),7(*S*)-bisepoxy-substituted compound **11** showed the most potent cell antiproliferative activity against all tested cancer cell lines with particularly strong inhibition of K562 myelogenous leukemia cancer cell proliferation.

4.2.2 Chiral hydroxyanthraquinone analogs

Natural and synthetic analogs of hydroxyanthraquinones (e.g., anthracyclines, mitoxantrone and emodin) are additional prototypes for the design of anticancer drug candidates with the ability to bind double-stranded DNA and inhibit topoisomerases 1 and 2 mediated relaxation of supercoiled DNA [90, 91].

Derivatives of (4,11-dihydroxynaphtho[2,3-*f*]indole-5,10-dione) were identified as a promising scaffold for the search of agents active against resistant tumor cells [92]. Shchekotikhin et al. [93] explored new TopI and TopII antagonists based on a 4,11-dihydroxy naphtho[2,3-*f*]indole-5,10-dione scaffold bearing the cyclic chiral diamine in the side chain - **Figure 16**.

Potent cytotoxicity (at submicromolar to low micromolar concentrations) against a panel of wild type mammalian tumor cells and isogenic drug resistant sublinies was observed for all novel derivatives of naphtho[2,3-*f*]indole-5,10-diones. Only isomer 7 induced the formation of specific DNA cleavage products similar for classical Top1 inhibitors camptothecin and indenoisoquinoline MJ-III-65. Importantly, the derivative of (*R*)- aminopyrrolidine 7 increased the life span of mice bearing P388 leukemia while its enantiomer **6** was inactive.

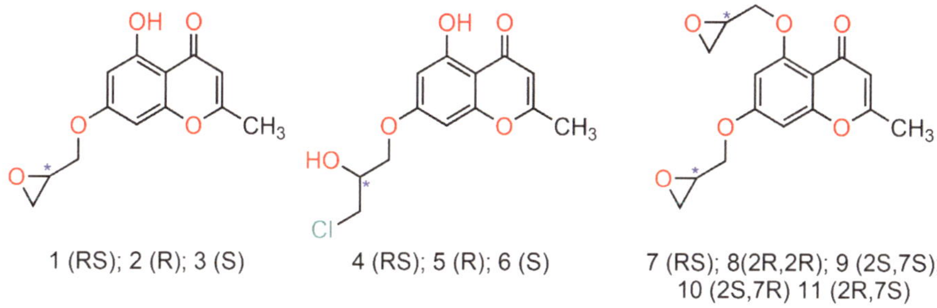

Figure 15.
The structures of prepared chiral chromane analogs.

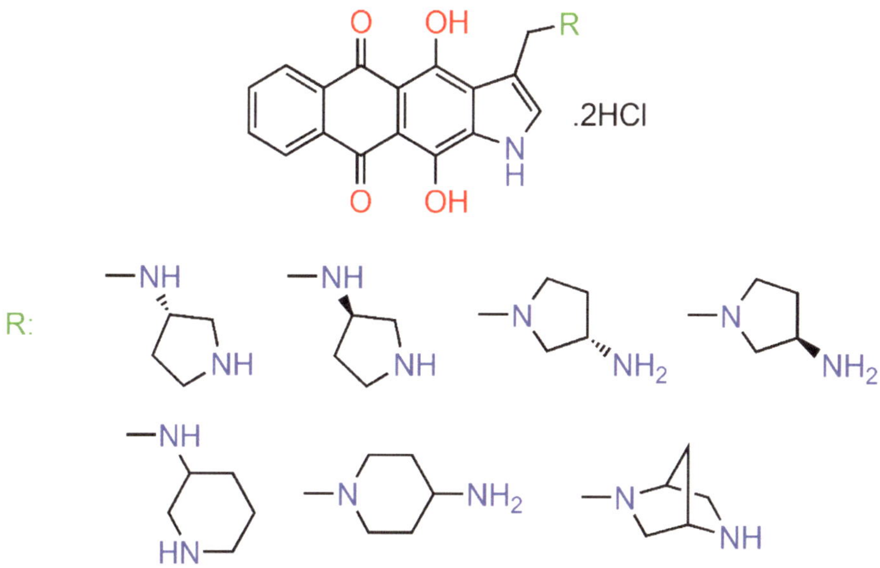

Figure 16.
The chemical structures of naphthoindolediones with some chiral substituents.

4.3 Chiral thiosemicarbazones

Another group of potent anticancer agents is made up from chiral thiosemicarbazones derived from homochiral amines **Figure 17** [94]. Their antiproliferative activity was evaluated against several panels of cancel cell lines (A549 (human alveolar adenocarcinoma), MCF-7 (human breast adenocarcinoma), HeLa (human cervical adenocarcinoma), and HGC-27 (human stomach carcinoma) cell lines. Some of the compounds, especialy thiosemicarbazones with substituted hydroxyl group, 4-chlorophenoxy, 4-fluorophenoxy showed inhibitory activities on the growth of cancer cell lines. Compounds with substituted piperidine ring exhibited higher activity against HCG-27 than taxol, which was uses as the standard. In every active thiosemicarbazone derivatives the (*S*)-enantiomer was more active than (*R*)-enantiomer. The most active compounds the (*S*)-isomers of the thiosemicarbazone derivative with substituted piperidine group, also showed the best fit for the generated pharmacophore hypothesis. In the pharmacophore model, the (*S*)-enantiomer was better matched than the (*R*)-enantiomer. This showed that the configuration of isomers greatly influenced their activity.

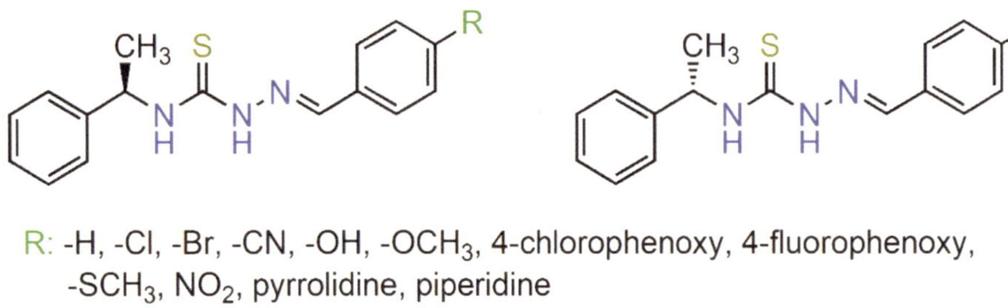

Figure 17.
The chemical structures of chiral thiosemicarbazones.

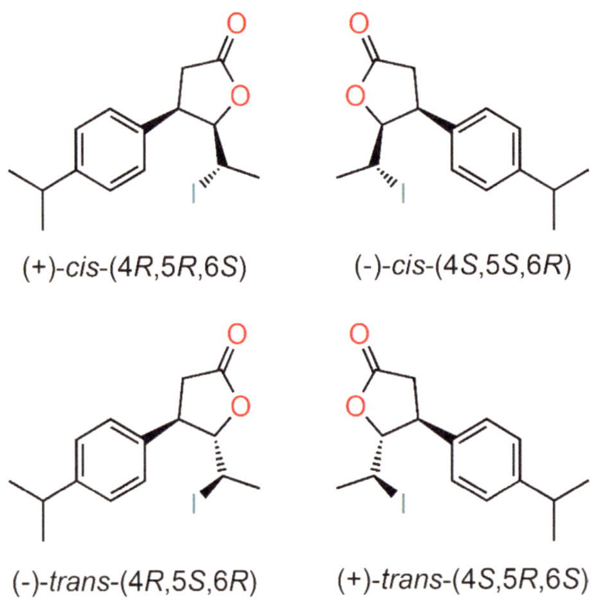

Figure 18.
The structure of stereoisomers δ-iodo-γ-lactones with p-isopropylphenyl substituent.

4.4 Chiral δ-iodo-γ-lactones

Compounds with both a lactone function and an aromatic ring in their structure are promising as potential anticancer agents [95, 96]. Four stereoisomers of δ-iodo-γ-lactones with a 4-isopropylphenyl substituent at the β-position - **Figure 18** were tested against a broad panel of canine cancer cell lines representing hematopoietic and mammary gland cancers. The investigated isomers exerted higher activity against canine lymphoma/leukemia cell lines than against mammary tumors, whereas the configuration of stereogenic centres of the examined compounds affected their activity.

Stereoisomers with the 4*S* configuration shown to be were more active, with the cis-(4*S*,5*S*,6*R*) isomer as the most potent. The investigated δ-iodo-γ-lactones act as an anticancer agent by the induction of apoptosis of canine cancer cells via a mitochondrial-mediated, caspase-dependent pathway [97].

5. Conclusions

Chirality has become a major task for the synthesis and development of drugs. One enantiomer of a chiral drug may be a medicine for particular disease whereas; another enantiomer of the molecule may be not only inactive but can even be toxic. The pharmacological activity of chiral drugs depends mainly on the drug's interaction with chiral biological molecules such as proteins, nucleic acids and bio membranes. This is valid not only for organic drugs, but also for metal-based drugs which are developed for anticancer application. This review outlines the effect of chirality on the efficiency of anticancer metal-based drugs and interesting potential organic chiral drug molecules. The influence of stereoselectivity on anticancer activity can hardly be generalized, it is manifested specifically for each individual chiral compound and depend on the type of cellular targets. However, knowledge of the stereochemistry of anticancer compounds can influence some critical processes underlying their toxicity towards cancer cells and provide a rational basis for the design of new antitumor drugs.

Acknowledgements

This work was supported by VEGA grant No 1/0145/20 (Slovak Republic).

Conflict of interest

The authors declare no conflict of interest.

Author details

Jindra Valentová* and Lucia Lintnerová
Department Chemical Theory of Drugs, Faculty of Pharmacy, Comenius University in Bratislava, Bratislava, Slovak Republic

*Address all correspondence to: valentova@fpharm.uniba.sk

References

[1] Patel BK, Hutt AJ. Stereoselectivity in Drug Action and Disposition: An Overview. In: Reddy I K, Mehvar R. Chirality in Drug Design and Development, 1st ed. Boca Raton: CRC Press; 2004. p. 139-190. DOI: 10.1201/9780203021811

[2] Lin G-Q, Zhang J-G, Cheng J-F. Overview of Chirality and Chiral Drugs. In: Lin. G-Q, You Q-D, Cheng J-F. Chiral Drugs: Chemistry and Biological Action. John Wiley & Sons; 2011. p. 5-21. DOI: 10.1002/9781118075647

[3] Hutt AJ. Chirality and pharmacokinetics: an area of neglected dimensionality? Drug Metabol Drug Interact. 2007;22(2-3):79-112. DOI: 0.1515/dmdi.2007.22.2-3.79.

[4] Agrawal YK, Bhatt HG, Raval HG, Oza PM, Gogoi PJ. Chirality – A New Era of Therapeutics. Mini-Reviews in Medicinal Chemistry. 2007;7:451-460. DOI: 10.2174/138955707780619617

[5] Čižmáriková R, Valentová J, Horáková R. Chirality of β_2-agonists. An overview of pharmacological activity, stereoselective analysis, and synthesis. *Open Chemistry. 2020; 18:* DOI: 10.1515/chem-2020-0056

[6] Hutt A, Valentová J. The chiral switch: the development of single enantiomer drugs from racemates. Acta Facultatis Pharmaceuticae Universitatis Comenianae. 2003;50:1-16.

[7] Gal J. Chiral drugs from a historical point of view. In: Francotte E, Linder W. Chirality in drug research. Weinheim: Wiley-VCH; 2006; p. 3-26. DOI: 10.1002/9783527609437.ch1

[8] Calcaterra A, D'Acquarica I. The market of chiral drugs: Chiral switches versus de novo enantiomerically pure compounds. Journal of Pharmaceutical and Biomedical Analysis. 2018;147:323-340 DOI: 10.1016/j.jpba.2017.07.008 .

[9] Wang Y, Huang H, Zhang Q, Zhang P. Chirality in metal-based anticancer agents. Dalton Transactions. 2018;47:4017-4026. DOI: 10.1039/c8dt00089a

[10] National Cancer Institute, Available from https://www.cancer.gov/about-cancer/understanding/what-is-cancer [Accessed 2021-01-12]

[11] Patrick GL. An Introduction to Medicinal Chemistry. 3rd ed. Oxford University Press; 2008.

[12] Health statistics and information systems. World Health Organization; 2021. Available from: https://www.who.int/health-topics/cancer#tab=tab_1 [Accessed 2021-01-12]

[13] Blagosklonny M V. Matching targets for selective cancer therapy. Drug Discovery Today. 2003;8:1104-1107. DOI: 10.1016/s1359-6446(03)02806-x

[14] Alama A, Orengo AM, Ferrini S, Gangemi R. Targeting cancer-initiating cell drug-resistance: a roadmap to a new-generation of cancer therapies? Drug Discovery Today. 2012;17:435-442. DOI: 10.1016/j.drudis.2011.02.005 .

[15] Dembic Z. Antitumor Drugs and Their Targets. Molecules. 2020;25:5776. DOI: 10.3390/molecules25235776

[16] Yousuf I, Bashir M. Metallodrugs in Medicine: Present, Past, and Future Prospects (Pages: 1-39) In: Shahid-ul-Islam Hashmi AA, Khan SA. Advances in Metallodrugs: Preparation and Applications in Medicinal Chemistry. Scrivener Publishing LLC; 2020. p. 1-40. DOI: 10.1002/9781119640868.ch1

[17] Frezza M, Hindo S, Chen D, Davenport A, Schmitt S, Tomco D, Dou QP. Novel Metals and Metal Complexes as Platforms for Cancer Therapy. Current Pharmaceutical

Design. 2010; 16:1813-1825. DOI: 10.2174/138161210791209009

[18] Ndagi U, Mhlongo N, Soliman ME. Metal complexes in cancer therapy – an update from drug design perspective. Drug Design, Development and Therapy. 2017;11:599-616. DOI: 10.2147/DDDT.S119488

[19] Nguyen LA, He H, Pham-Huy C. Chiral drugs: an overview. International Journal of Biomedical Science. 2006;2:85-100

[20] Arnesano F, Pannunzio A, Coluccia M, Natile G. Effect of chirality in platinum drugs. Coordination Chemistry Reviews. 2015; 284:286-297. DOI: 10.1016/j.ccr.2014.07.016

[21] Romero MJ, Sadler PJ. Chirality in Organometallic Anticancer Complexes. In: Jaouen G, Salmain M. Bioorganometallic Chemistry. Weinheim: Wiley-VCH; 2014. p. 85-116 . DOI: 10.1002/9783527673438

[22] Conelly NG, Damhus T, Hartshorn RM, Hutton AT. Nomenclature of Inorganic Chemistry: IUPAC Recommendations. RSC Publishing, London; 2005. DOI: 10.1515/pac-2014-0718

[23] Dilruba S, Kalayda GV. Platinum-based drugs: past, present and future. Cancer Chemotherapy and Pharmacology. 2016;77:1103-24. 10.1007/s00280-016-2976-z

[24] Wheate NJ, Walker S, Craig GE, Oun R. The status of platinum anticancer drugs in the clinic and in clinical trials. Dalton Transactions. 2010;39:8113-8127. DOI: 10.1039/C0DT00292E

[25] Galanski M, Jakupec MA, Keppler BK. Update of the Preclinical Situation of Anticancer Platinum Complexes: Novel Design Strategies and Innovative Analytical Approaches. Current Medicinal Chemistry. 2005;12:2075-2094. DOI: 10.2174/0929867054637626

[26] Jansen BAJ, Brouwer J, Reedijk J. Glutathione Induces Cellular Resistance Against Cationic Dinuclear Platinum Anticancer Drugs. Journal of Inorganic Biochemistry. 2002;89:197-202. DOI: 10.1016/s0162-0134(02)00381-1

[27] Farrell NP. Progress in Platinum-derived Drug Development. Drugs Future, 2012;37:795-806. DOI: 10.1358/dof.2012.037.011.1830167

[28] Johnstone TC, Suntharalingam K, Lippard SJ. The Next Generation of Platinum Drugs: Targeted Pt(II) Agents, Nanoparticle Delivery, and Pt(IV) Prodrugs. Chemical Reviews. 2016;116:3436-3486. DOI: 10.1021/acs.chemrev.5b00597

[29] Varbanov HP, Goeschl S, Heffeter P, Theiner S, Roller A, Jensen F, Jakupec MA, Berger W, Galanski M, Keppler BKA. Novel Class of Bis- and Tris-ChelateDiam(m)ine Bis(dicarboxylato)Platinum(IV) Complexes as Potential Anticancer Prodrugs. Journal of Medicinal Chemistry. 2014;57:6751-6764. DOI: 10.1021/jm500791c

[30] Abu-Surrah AS, Kettunen M. Platinum Group Antitumor Chemistry: Design and development of New Anticancer Drugs Complementary to Cisplatin. Current Medicinal Chemistry. 2006;13:1337-1357. DOI: 10.2174/092986706776872970

[31] Hanif M, Hartinger CG. Anticancer metallodrugs: where is the next cisplatin? Future Medicinal Chemistry. 2018;10:615-617. DOI: 10.4155/fmc-2017-0317

[32] Ghosh S. Cisplatin: The first metal based anticancer drug. Bioorganic Chemistry. 2019;88:102925. DOI: 10.1016/j.bioorg.2019.102925

[33] Kim D-K, Kim G, Gam J, Cho Y-B, Kim H-T, Tai J-H, Kim KH, Hong W-S, Park J-G. Synthesis and antitumor activity of a series of [2-substituted-4,5-bis(aminomethyl)-1,3-dioxolane] Pt(II) complexes. Journal of Medicinal Chemistry. 1994;37:1471-85. DOI: 10.1021/jm00036a013

[34] Fuertes M, Castilla J, Alonso C, Pérez J. Cisplatin Biochemical Mechanism of Action: From Cytotoxicity to Induction of Cell Death Through Interconnections Between Apoptotic and Necrotic Pathways. Current Medicinal Chemistry, 2003;10:257-266. DOI: 10.2174/0929867033368484

[35] Dasari S, Tchounwou PB. Cisplatin in cancer therapy: molecular mechanisms of action. European Journal of Pharmacology. 2014;740:364-78. DOI: 10.1016/j.ejphar.2014.07.025

[36] Benedetti M, Malina J, Kasparkova J, Brabec V, Natile G. Chiral discrimination in platinum anticancer drugs. Environmental Health Perspectives. 2002;110:779-82. DOI: 10.1289/ehp.02110s5779

[37] Hambley TW. The influence of structure on the activity and toxicity of Pt anti-cancer drugs. Coordination Chemistry Reviews. 1997;166:181-223. DOI: 10.1016/S0010-8545(97)00023-4

[38] Inagaki K, Kidani Y. Differences in binding of (1,2-cyclohexanediamine) Pt(II) isomers with d(GpG). Inorganic Chemistry. 1986;25:1-3. DOI: 10.1021/ic00221a001

[39] Abramkin SA, Jungwirth U, Valiahdi SM, Dworak C, Habala L, Meelich K, Berger W, Jakupec MA, Hartinger CG, Nazarov AA, Galanski M, Keppler BKA. {(1*R*,2*R*,4*R*)-4-methyl-1,2-cyclohexanediamine} oxalatoplatinum(II): A Novel Enantiomerically Pure Oxaliplatin Derivative Showing Improved Anticancer Activity in Vivo. Journal of Medicinal Chemistry. 2010;53:7356-7364. DOI: 10.1021/jm100953c

[40] Liu F, Gou S, Chen F, Fang L, Zhao J. Study on Antitumor Platinum(II) Complexes of Chiral Diamines with Dicyclic Species as Steric Hindrance. Journal of Medicinal Chemistry Med Chem. 2015;58:6368-6377. DOI: 10.1021/jm501952r

[41] Dufrasne F, Galanski M. The Relation Between Stereochemistry and Biological Activity of Platinum(II) Complexes Chelated with Chiral Diamine Ligands: An Intricate Problem. Current Pharmaceutical Design. 2007;13:2781-2794, DOI: 10.2174/138161207781757060

[42] Wappes B, Jennerwein M, von Angerer E, Schönenberger H, Engel J, Berger M, Wrobel KH. Dichloro-[1,2-bis-(4-hydroxyphenyl) ethylenediamine] platinum(II) complexes: an approach to develop compounds with a specific effect on the hormone-dependent mammary carcinoma. Journal of Medicinal Chemistry. 1984;27:1280-6. DOI: 10.1021/jm00376a009

[43] Bernhardt G, Gust R, Reile H, vom Orde HD, Müller R, Keller C, Spruß T, Schönenberger H, Burgemeister T, Mannschreck A, Range KJ, Klement U. [1,2-Bis(2-hydroxyphenyl) ethylenediamine]dichloroplatinum (II), a new compound for the therapy of ovarian cancer. Journal of Cancer Research and Clinical Oncology. 1992;118:209-15. DOI: 10.1007/BF01410136

[44] Lindauer E, Holler E. Cellular distribution and cellular reactivity of platinum (II) complexes. Biochemical Pharmacology. 1996;52:7-14. DOI: 10.1016/0006-2952(96)00106-2

[45] Morikawa K, Honda M, Endoh K-I, Matsumoto T, Akamatsu K-I,Mitsui H,

Koizumi M. Synthesis, antitumor activity, and nephrotoxicity of the optical isomers of 2-aminomethylpyrrolidine(1,1-cyclobutanedicarboxylato) platinum(II). Journal of Pharmaceutical Sciences. 1991; 80:837-42. DOI: 10.1002/jps.2600800907

[46] Sekaran V, Soares J, Jarstfer MB. Telomere maintenance as a target for drug discovery. Journal of Medicinal Chemistry. 2014;57:521-538. DOI: 10.1021/jm400528t

[47] Wu RA, Collins K. Sequence specificity of human telomerase. Proceedings of the National Academy of Sciences of the United States of America. 2014;111:11234-11235. DOI: 10.1073/pnas.1411276111

[48] Li Z, Liu C, Huang C, Meng X, Zhang L, He J, Li J. Quinazoline derivative QPB-15e stabilizes the c-myc promoter G-quadruplex and inhibits tumor growth *in vivo*. Oncotarget. 2016;7:21658-21675. DOI: 10.18632/oncotarget.9088

[49] Xiong YX, Su HF, Lv P, Ma Y, Wang SK, Miao H, Liu HY, Tan JH, Ou TM, Gu LQ, Huang ZS. A newly identified berberine derivative induces cancer cell senescence by stabilizing endogenous G-quadruplexes and sparking a DNA damage response at the telomere region. Oncotarget. 2015,6:35625-35635. DOI: 10.18632/oncotarget.5521

[50] Zhang J, Zhang F, Li H, Liu C, Xia J, Ma L, Chu W, Zhang Z, Chen C, Li S, Wang S. Recent progress and future potential for metal complexes as anticancer drugs targeting G-quadruplex DNA. Current Medicinal Chemistry. 2012;19:2957-75. DOI: 10.2174/092986712800672067

[51] Carvalho J, Mergny JL, Salgado GF, Queiroz JA, Cruz C. G-quadruplex, Friend or Foe: The Role of the G-quartet in Anticancer Strategies. Trends in Molecular Medicine. 2020;26:848-861. DOI: 10.1016/j.molmed.2020.05.002

[52] Qin Q-P, Qin J-L, Chen M, Li Y-L, Meng T, Zhou J, Liang **H, C**hen Z-F. Chiral platinum (II)-4-(2,3-dihydroxypropyl)-formamide oxo-aporphine (FOA) complexes promote tumor cells apoptosis by directly targeting G-quadruplex DNA *in vitro* and *in vivo.* **Oncotarget, 2017;8: 61982-61997. D**OI: 10.18632/oncotarget.18778

[53] Chen Z-F, Qin Q-P, Qin J-L, Zhou J, Li Y-L, Li N, Liu Y-C, Liang H. Water-Soluble Ruthenium(II) Complexes with Chiral 4-(2,3-dihydroxypropyl)-formamide oxoaporphine (FOA): in Vitro and in Vivo Anticancer Activity by Stabilization of G-Quadruplex DNA,Inhibition of Telomerase Activity and Induction of Tumor Cell Apoptosis. Medicinal Chemistry. 2015;58:4771-89. DOI: 10.1021/acs.jmedchem.5b00444

[54] Pal M, Nandi U, Mukherjee D. Detailed account on activation mechanisms of ruthenium coordination complexes and their role as antineoplastic agents. European Journal of Medicinal Chemistry. 2018;150:419-445. DOI: 10.1016/j.ejmech.2018.03.015

[55] Marloye M, Berger G, Gelbcke M, Dufrasne F. A survey of the mechanisms of action of anticancer transition metal complexes. Future Medicinal Chemistry. 2016;8:2263-2286. DOI: 10.4155/fmc-2016-0153

[56] Zeng L, Gupta P, Chen Y, Wang E, Ji L, Chao H, Chen ZS. The development of anticancer ruthenium(ii) complexes: from single molecule compounds to nanomaterials. Chemical Society Reviews. 2017;46:5771-5804. DOI: 10.1039/c7cs00195a

[57] Bergamo, A.; Sava, G. Ruthenium anticancer compounds: myths and

realities of the emerging metal-based drugs. Dalton Transactions. 2011;40:7817–7823. DOI: 10.1039/C0DT01816C

[58] Brabec V, Novakova O. DNA binding mode of ruthenium complexes and relationship to tumor cell toxicity. Drug Resistance Updates. 2006;9:111-122. DOI: 10.1016/j.drup.2006.05.002

[59] Lee SY, Kim CY, Nam T-G. Ruthenium Complexes as Anticancer Agents: A Brief History and Perspectives. Drug Design, Development and Therapy. 2020;14:5375-5392. DOI: 10.2147/DDDT.S275007

[60] Wang J-Q, Zhang P-Y, Qian C, Hou X-J, Ji L-N, Chao H. Mitochondria are the primary target in the induction of apoptosis by chiral ruthenium(II) polypyridyl complexes in cancer cells. Journal of Biological Inorganic Chemistry. 2014;19:335-348. DOI: 10.1007/s00775-013-1069-2

[61] Liao G, Chen X, Wu J, Qian C, Wang H, Ji L, Chao H. Novel ruthenium(II) polypyridyl complexes as G-quadruplex stabilisers and telomerase inhibitors. Dalton Transactions. 2014;43:7811-7819. DOI: 10.1039/C3DT53547A

[62] Yu Q, Liu Y, Zhang J, Yang F, Sun D, Liu D, Zhou Y, Liu J. Ruthenium(II) polypyridyl complexes as G-quadruplex inducing and stabilizing ligands in telomeric DNA. Metallomics. 2013;5:222-231. DOI: 10.1021/acs.jmedchem.7b01689

[63] Sun D, Liu Y, Yu Q, Liu D, Zhou Y, Liu J. Selective nuclei accumulation of ruthenium(II) complex enantiomers that target G-quadruplex DNA, Journal of Inorganic Biochemistry. 2015;150: 90-99. DOI: 10.1016/j.jinorgbio.2015.04.003

[64] Thota S, Rodrigues DA, Crans DC, Barreiro EJ. Ru(II) Compounds: Next-Generation Anticancer Metallotherapeutics? Journal of Medicinal Chemistry. 2018;61:5805-5821. DOI: 10.1021/acs.jmedchem.7b01689

[65] Atilla-Gokcumen GE, Williams DS, Bregman H, Pagano N, Meggers E. Organometallic compounds with biological activity: A very selective and highly potent cellular inhibitor for glycogen synthase kinase 3. ChemBioChem. 2006;7:1443-1450. DOI: 10.1002/cbic.200600117

[66] Carter R, Westhorpe A, Romero MJ, Habtemariam A, Galleevo CR, Bark Y, Menezes N, Sadler PJ, Sharma RA. Radiosensitisation of human colorectal cancer cells by ruthenium(II) arene anticancer complexes. Scientific Reports. 2016;6:20596. DOI: 10.1038/srep20596

[67] de la Cueva-Alique I, Sierra S, Muñoz-Moreno L, Pérez-Redondo A, Bajo AM, Marzo I, Gude L, Cuenca T, Royo E. Biological evaluation of water soluble Arene Ru(II) enantiomers with aminooxime. Journal of Inorganic Biochemistry. 2018;183: 32-42. DOI: 10.1016/j.jinorgbio.2018.02.018

[68] Fu Y, Soni R, Romero MJ, Pizarro AM, Salassa L, Clarkson GJ, Hearn JM, Habtemariam A, Wills M, Sadler PJ. Mirror-Image Organometallic Osmium Arene Iminopyridine Halido Complexes Exhibit Similar Potent Anticancer Activity. Chemistry – A European Journal. 2013;19:15199-15209. DOI: 10.1002/chem.201302183

[69] Chen L-A, Ding X, Gong L, Meggers E. Thioether-based anchimeric assistance for asymmetriccoordination chemistry with ruthenium(II) and osmium(II). Dalton Transactions. 2013;42:5623-5626. DOI: 10.1039/C3DT00015J

[70] Göbel P, Ritterbusch F, Helms M, Bischof M, Harms K, Jung M,

Meggers E. Probing chiral recognition of enzyme active sites with octahedral iridium(III) propeller complexes. European Journal of Inorganic Chemistry. 2015;10:1654-1659. DOI: 10.1002/ejic.201500087

[71] Kang T-S, Mao Z, Ng C-T, Wang M, Wang W, Wang C, Lee SM-Y, Wang Y, Leung C-H, Ma D-L. Identification of an iridium(III)-based inhibitor of tumor necrosis factor. Journal of Medicinal Chemistry. 2016;59:4026-4031. DOI: 10.1021/acs.jmedchem.6b00112

[72] Rajaratnam R, Martin EK, Dörr M, Harms K, Casini A, Meggers E. Correlation between the Stereochemistry and Bioactivity in Octahedral Rhodium Prolinato Complexes. Inorganic Chemistry. 2015;54:8111-8120. DOI: 10.1021/acs.inorgchem.5b01349

[73] Mullick AB, Chang YM, Ghiviriga I, Abboud KA, Tan W, Veige AS. Human cancerous and healthy cell cytotoxicity studies of a chiral μ-dicarbene–digold (I) metallamacrocycle. Dalton Transactions. 2013;42:7440-7446. DOI: 10.1039/C3DT32844A

[74] Bertrand B, Williams MRM, Bochmann M. Gold(III) Complexes for Anti-tumour Applications: an Overview. Chemistry A European Journal. 2018;24:11840-11851. DOI: 10.1002/chem.201800981

[75] Mora M, Gimeno MC, Visbal R. Recent advances in gold-NHC complexes with biological properties. Chemical Society Reviews. 2019;48:447-462. DOI: 10.1039/C8CS00570B

[76] Li BB, Jia YX, Zhu PC, Chew RJ, Li Y, Tan NS, Leung PH. Highly selective anti-cancer properties of ester functionalized enantiopure dinuclear gold(I)-diphosphine. European Journal of Medicinal Chemistry. 2015;98:250-255. DOI: 10.1016/j.ejmech.2015.05.027

[77] Song Y, Vittal JJ, Srinivasan N, Chan S-H, Leung P-H. Synthesis and anti-cancer activities of a pair of enantiomeric gold (I) complexes containing sulfanyl-substituted P-stereogenic phosphines. Tetrahedron: Asymmetry. 1999;10:1433-1436. DOI: 10.1016/S0957-4166(99)00106-8

[78] Boselli L, Ader I, Carraz M, Hemmert C, Cuvillier O, Gornitzka H. Synthesis, structures, and selective toxicity to cancer cells of gold(I) compexes involving N-heterocyclic carbene ligands. European Journal of Medicinal Chemistry. 2014;85:87-94. DOI: 10.1016/j.ejmech.2014.07.086

[79] Zou TT, Lum CT, Lok CN, Zhang JJ, Che CM. Chemical biology of anticancer gold(III) and gold(I) complexes. Chemical Society Reviews. 2015;44:8786—8801. DOI: 10.1039/C5CS00132C

[80] Anastasia De Luca A, Hartinger ACH, Dyson PJ, Lo Bello M, Casini A. A new target for gold(I) compounds: glutathione-S-transferase inhibition by auranofin J Inorg Biochem . 2013, 119, 38-42. DOI: 10.1016/j.jinorgbio.2012.08.006

[81] Arojojoye AS, Mertens RT, Ofori S, Parkin SR, Awuah SG. Synthesis, Characterization, and Antiproliferative Activity of Novel Chiral $[QuinoxP^{*}AuCl_2]^{+}$ Complexes. Molecules. 2020; 25: 5735; DOI: 10.3390/molecules25235735

[82] Armando GR, Mengual Gómez DL, Gomez DE. New drugs are not enough-drug repositioning in oncology: An update. International Journal of Oncology. 2020;56:651-684. DOI: 10.3892/ijo.2020.4966

[83] Kim K-W, Roh JK, Wee H-J, Kim C. Alkylating Anticancer Drugs. In: Kim K-W, Roh JK, Wee H-J, Kim C. Cancer Drug Discovery. Dordrecht: Springer;

2016. p. 71-94. DOI: 10.1007/978-94-024-0844-7_4

[84] Rajski SR, Williams RM. DNA Cross-Linking Agents as Antitumor Drugs. Chemical Reviews. 1998;98: 2723–2796. DOI: 10.1021/cr9800199

[85] Sosic A, Zuravka I, Schmitt NK, Miola A, Gottlich R, Fabris D, Gatto B. Direct and Topoisomerase II Mediated DNA Damage by Bis-3-chloropiperidines: The Importance of Being an Earnest G. ChemMedChem. 2017;12:1471–1479. DOI: 10.1002/cmdc.201700368

[86] Zuravka I, Roesmann R, Sosic A, Gottlich R, Gatto B. Bis-3-chloropiperidines containing bridging lysine linkers: Influence of side chain structure on DNA alkylating activity. Bioorganic Medicinal Chemistry. 2015;23:1241–1250. DOI: 10.1016/j.bmc.2015.01.050

[87] Carraro C, Francke A, Sosic A, KohlF, Helbing, T, De Franco M, Fabris D, Göttlich R, Gatto B. Behind the Mirror: Chirality Tunes the Reactivity and Cytotoxicity of Chloropiperidines as Potential Anticancer Agents. ACS Med. Chem. Lett. 2019, 10, 552–557 DOI:10.1021/acsmedchemlett.8b00580

[88] Pommier Y, Leo E, Zhang H, Marchand C. DNA topoisomerases and their poisoning by anticancer and antibacterial drugs. Chemistry & Biology. 2010;17:421-433. DOI: 10.1016/j.chembiol.2010.04.012

[89] Jo H, Seo SH, Na Y, Kwona Y. The synthesis and anticancer activities of chiral epoxy-substituted chromone analogs. Bioorganic Chemistry. 2019;84:347-354. DOI: 10.1016/j.bioorg.2018.11.054

[90] Chiang J-H, Yang J-S, Ma C-Y, Yang M-D, Huang H-Y, Hsia T-C, Kuo H-M, Wu P-P,Lee T-H, Chung J-G. Danthron, an anthraquinone derivative, induces DNA damage and caspase cascades-mediated apoptosis in SNU-1 human gastric cancer cells through mitochondrial permeability transition pores and Bax-triggered pathways. Chemical Research in Toxicology. 2011;24:20-29. DOI: 10.1021/tx100248s

[91] Zeng G-Z, Fan J-T, Xu J-J, Li Y, Tan N-H. Apoptosis induction and G2/M arrest of 2-methyl-1,3,6-trihydroxy-9,10-anthraquinone from *Rubia yunnanensis* in human cervical cancer HeLa cells. Die Pharmazie. 2013;68:293-299. DOI: 10.1691/ph.2013.2808

[92] Shchekotikhin AE, Shtil AA, Luzikov YN, Bobrysheva TV, Buyanov VN, Preobrazhenskaya MN. 3-Aminomethyl derivatives of 4,11-dihydroxynaphtho[2,3-*f*]indole-5,10-dionefor circumvention of anticancer drug resistance. Bioorganic and Medicinal Chemistry. 2005;13:2285-2291. DOI: 10.1016/j.bmc.2004.12.044

[93] Shchekotikhin AE, Glazunova VA, Dezhenkova LG, Luzikov YN, Buyanov VN, Treshalina HM, Lesnaya NA, Romanenko VI, Kaluzhny DN, Balzarini J, Agama K, Pommier Y, Shtil AA, Preobrazhenskaya MN. Synthesis and evaluation of new antitumor 3-aminomethyl-4,11-dihydroxynaphtho[2,3-*f*]indole-5,10-diones. European Journal of Medicinal Chemistry. 2014;86:797-805. DOI: 10.1016/j.ejmech.2014.09.021

[94] Taşdemir D, Karaküçük-İyidoğan A, Ulaşli M, Taşkin-Tok T, Oruç-Emre EE, Bayram H. Synthesis, Molecular Modeling, and Biological Evaluation of Novel Chiral Thiosemicarbazone Derivatives as Potent Anticancer Agents. Chirality. 2015;27:177-188. DOI: 10.1002/chir.22408

[95] Gładkowski W, et al., Chiral δ-iodo-γ-lactones derived from cuminaldehyde, 2, 5-dimethylbenzaldehyde and

piperonal: chemoenzymatic synthesis and antiproliferative activity. Tetrahedron: Asymmetry. 2016;27: 227-237. DOI: 10.1016/j.tetasy.2016.02.003

[96] Albrecht A, Koszuk JF, Modranka J, Rózalski M, Krajewska U, Janecka A, Studzian K, Janecki T. Synthesis and cytotoxic activity of gamma-aryl substituted alpha-alkylidene-gamma-lactones and alpha-alkylidene-gamma-lactams. Bioorganic Medicinal Chemistry. 2008;16:4872-4882. DOI: 10.1016/j.bmc.2008.03.035

[97] Pawlak A, Gładkowski W, Mazur M, Henklewska M, Obmińska-Mrukowicz B, Rapak A. Optically active stereoisomers of 5-(1-iodoethyl)-4-(40-isopropylphenyl) dihydrofuran-2-one: The effect of the configuration of stereocenters on apoptosis induction in canine cancer cell lines. Chemico-Biological Interactions. 2017;261:18-26. DOI: 10.1016/j.cbi.2016.11.013

Chapter 9

Mirror Symmetry of Life

Beata Zagórska-Marek

Abstract

Functioning in the Earth gravity field imposes on living organisms a necessity to read directions. The characteristic feature of their bodies, regardless unicellular or multicellular, is axial symmetry. The development of body plan orchestrated by spatiotemporal changes in gene expression patterns is based on formation of the vertical and radial axes. Especially for immobile plants, anchored to the substrate, vertical axis is primary and most important. But also in animals the primary is the axis, which defines the anterior and posterior pole of the embryo. There are many little known chiral processes and structures that are left- or right oriented with respect to this axis. Recent developments indicate the role of intrinsic cell chirality that determines the direction of developmental chiral processes in living organisms. The still enigmatic events in cambia of trees and handedness of phyllotaxis as well as plant living crystals are in focus of the chapter.

Keywords: intrinsic cell chirality, charophytes, cambium, morphogenesis, figured wood, plant development, phyllotaxis, shoot apical meristem, snail shells, handedness, dislocations

I see the mirror fairy tale of infinite reflections spinning out to weave no end....
Bolesław Leśmian "Prolog"

1. Introduction

Polarized environment, from the beginning of life on our planet, imposes on living organisms a necessity to read directions. Light and gravity are the two most important oriented signals that come from well-defined sources. The response to these primary polar signals allowed for development of the secondary, more sophisticated reactions to the network of other polar signals like gradients of chemical molecules or mechanical stresses. The signals of chemical and physical nature provide the extrinsic but also an intrinsic information for biological systems.

One of the basic features of the organism developing in polar environment is its axiality. This brings in consequence following possibilities: 1) multiplication of repetitive units of the body and their special alignment along the axis (segmentation, metamerism) and 2) deviations of structures from the axis to the left or to the right (development of L/R symmetry).

The main axis in the motionless plants, fastened to the ground, is mostly vertical, extending between the apical and basal poles. The animal main axis, regardless its position in the gravity field, connects the anterior and posterior poles. The axis

IntechOpen

formation starts at the very early stages of development. The identity of segments formed iteratively along the axis in both plants and animals is genetically controlled and their evolutionary multiplication creates a great potential for morphological and functional diversity through many useful modifications. This process is in a sense similar to the effects of gene duplication on the molecular level.

Subsequent emergence of L/R symmetry may be observed on all, hierarchically different levels of body organization. Some general principles, like minimum energy rule, are universal in nature leading to the identical solutions on all these levels. Good example represent spherical geodesic shapes. The structure of carbon allotrope - C_{60} closed fullerene, is also present in coated endocytic vesicles reinforced by the clathrin cage [1], in regularly sculptured surface of pollen grains [2] and in a cellular pattern on the surface of the plant paraboloid apical meristem [3].

Other universal basic forms are chiral helices and spirals commonly observed on molecular, cellular and organismal levels. They are of particular interest here because of distinct mirror symmetry they have, which is the main focus of this chapter. Chirality of many macromolecules: nucleic acids, proteins or cellulose fibers [4], coiled coils of collagen, or such structures as tubulin cytoskeleton, thickenings of plant cell wall, plant tendrils, spiral snail shells or narwhal tusks are but a few examples. Not only structures but also some developmental processes may be chiral. The apical cell divisions in moss gametophores [5], cell cleavage in the embryos of snails [6, 7] or lateral organ initiation on plant shoot apical meristem (SAM) proceed clockwise (CW) or counterclockwise (CCW). Two interesting problems may be addressed while considering chiral structures in biological systems – mechanism of their formation and proportion between the two chiral configurations.

The aim of this chapter is to provide the readers with the overview of some examples of bio-chirality discovered over the years both in animals and plants. The stronger accent will be placed on the latter because they are less known and because they have always been in a focus of the author's research. The mechanisms of many cases of mirror-symmetry presence in plants are yet to be elucidated.

2. Mirror symmetry on cellular level

Cell chirality or handedness is a newly discovered phenomenon, which nowadays is intensely studied, mostly in animal cells [7–10]. It is manifested in the presence of chiral structures within the cells but also in the cell behavior that may lead to directional movements or assuming L or R orientation of cell alignment. In animals it affects organs laterality [10], in plants results in development of spiral, helical or wavy patterns [3, 11–15].

During primary axis formation on the cellular level the polarity of the cell is manifested in an uneven distribution of receptors, ion channels and hormone carriers on plasma membrane, and internally in the ion currents and cytoskeletal fibers parallel to the developing axis but also in the polar distribution of ultrastructural components like cell organelles or nutrients. All these sophisticated processes have been investigated mainly in plant egg cells or fucoid zygotes [16–18] and animal oocytes [19, 20]. However, even in the integrated system of multicellular organism, singular cell polarity is often a case. In animal body the epithelial cells of intestines constituting a planar 2D barrier, have their polarity unified. It is manifested in nonrandom distribution of glucose transporters which facilitates the oriented transepithelial sugar transport [21]. In plants the polar distribution of the auxin influx (AUX) and efflux (PIN) carriers in plasma membrane results in polar transport of the hormone between the cells [22]. In L1 layer of SAM auxin is transported acropetally, whereas inside of the plant body, in the provascular tissues and later in cambium, the hormone transport is basipetal. Change in the distribution of carriers and thus of the cell polarity redirects the transport, often affecting the directions of plant organ growth [23]. This, for instance, has been noted in gravitropic response of the roots [24].

Many elements of the cell ultrastructure are spectacularly chiral. In some green algae exemplified by multicellular filamentous *Spirogyra* or unicellular *Spirotaenia* and *Chlamydomonas spirogyroides* the chloroplasts are of considerable length, flat and ribbon-like. They assume helical course in the cortical cytoplasm of the cell. Not much is known about the chiral configurations of their coiling. Images available in various data bases suggest that in *Spirogyra* both configurations may be present in different filaments [25], in different cells of the same filament or even within the same cell [26]. However, the error resulting from improper focusing during microscopic observations cannot be excluded. The sample taken for analysis from the aquarium of the Botanical Garden of Wrocław showed under light microscope hundreds of cells of the same S configuration of chloroplasts coiling from the right to the left (**Figure 1**). Mechanisms by which the configuration is regulated remain undiscovered.

Another clearly chiral component of the cell is basal body. In eukaryotic, plant and animal cells some identical structures bear different names although they look the same. Two centrioles of the centrosome, basal bodies or kinetosomes in the motile or ciliary epithelial cells have the same architecture. Composed of 9 triplets of microtubules (MT), overlapping on all available images either CW or CCW, they resemble a pinwheel toy. It is unclear, however, whether both configurations, being a mirror-image of one another, are indeed present in all different types of the cells. Transmission electron micrographs (TEM) of tangentially sectioned cell surface show that in a particular cell all basal bodies underlying cilia are of the same chirality [27]. However, unless it is clearly stated like in [28], it is not known whether

Figure 1.
The same filament of Spirogyra *photographed under fluorescent microscope at different optical levels: at its upper surface (left) and well below, close to its opposite side (right). S helix of the chloroplast, visualized here by the red autofuorescence of chlorophyll, may be falsely interpreted as Z, when due to the changing focus is watched from the inside of the cell.*

basal body is seen on TEM from the surface of the cell or from its inside. This is the reason for chiral configurations of basal bodies being uncertain. The image of *Paramaecium micronucleatum* by Dennis Kunkel [27] shows CCW overlapping triplets, whereas in *Paramecium tetraurelia* [29] the triplets overlap CW.

In flagellar apparatus of *Chlamydomonas* or *Acrosiphonia* gamete both chiral configurations of basal bodies are present on the same electron micrograph [30, 31] but in these cases it is certainly an effect of opposite orientation of flagella and their two basal bodies facing each other horizontally. The bodies are in fact of the same chirality. This case shows again how careful one must be analyzing and interpreting the examples of mirror-symmetry of the chiral ultrastructural components of the cell on TEM images, that can be easily flipped and show the same structure either from above or from the bottom side. There is one additional aspect of mirror symmetry presence in the locomotion apparatus of green algae. The flagellar roots of the two basal bodies are slightly rotated one with respect to the other – CW in representatives of Chlorophyceae and CCW in Ulvophyceae [30, 32, 33]. This finding, among the others, was the foundation of the profound revision of green algae taxonomy [30, 33].

The process of cilia beating is chiral. Both ciliates, motile spermatozoids and stationary epithelial cells readily change the direction of cilia movement either to navigate or alter the current of surrounding fluids [34]. Interestingly, the latter has been employed by unicellular *Stentor rosei*, to avoid, in quite deliberate and calculated manner, the irritating particles experimentally added to the medium [35]. The cell, however, is not always in full control over the cilia beating. The doublets of some ciliates, having notably the same chirality of basal bodies, show that in the form being a mirror image of the typical one, food particles are expelled from the oral apparatus instead of being directed towards it [28]. Typical chiral form of *Paramaecium* swims forward employing leftward rotation. Stressful conditions like low temperature or heavy metals force the ciliate to spin in opposite direction [36].

Among fibrillar elements constituting cytoskeleton, a tensegral structure of eukaryotic cytoplasm it is actin microfilament (MF) that is chiral. Mammalian cells exhibit specific, actin dependent L/R asymmetry which is different in normal and cancerous cells and changes when inhibitors of actin function are applied [37]. In bacteria changeable chirality of actin homologs MreB has impact on their growth and cell shape [38]. Actin is also responsible for directional, chiral movement of cytoplasm in the cells of charophytes [39]. Another important component of cytoskeleton, MTs. *per se* are not chiral. However, their arrangement in the cortical cytoplasm, beneath plasma membrane, is often oriented in plant cells (**Figure 2**). The elongating cells in the axial organs of the model plant *Arabidopsis thaliana* apparently have the chirality of their cortical cytoskeleton genetically controlled. In *spiral 1* and *spiral 2* mutants the cortical MTs, watched from the cell surface, form an ascending S helix, whereas in *lefty* mutant the helix is Z. The mutations lead to abnormal growth of the plant axial organs which, strangely enough, become twisted oppositely to the configuration of MT helix in their cells – Z in the *spiral*, S in the *lefty* mutants [40–42].

Configuration of cortical cytoskeleton, below plasma membrane, in some differentiating plant cells may change with time causing the development of interlocked pattern of the cellulose microfibrils deposited in the secondary wall [43]. Both chiral systems: intracellular and extracellular, parallel each other in consecutive stages of the secondary cell wall formation. The cycle of changes in microfibrils orientation always starts from the ascending S helix in S1 layer of the secondary wall (watched from the outside of the cell), it alters to Z in the S2 layer and returns to S helix in the S3 layer. No exception from this rule has been found

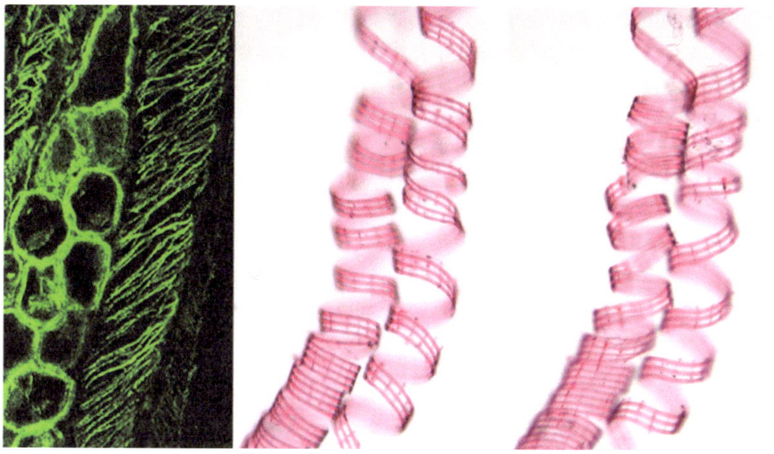

Figure 2.
Immunofluorescent visualization of Z helical cortical cytoskeleton in the elongated cambial cell of Cinnamomum camphora *(left) and two S helices of the secondary cell wall thickenings isolated from the protoxylem cells of* Scindapsus *sp. photographed at two different optical levels. The middle photo shows the surface of the thickenings seen from their outside.*

even though, at least theoretically, the opposite sequence of changes in microfibrils orientation is possible. Moreover, the cycle of changes in the chirality of microfibrillar helix appears to be independent of the overall orientation of the differentiating cell within figured wood, which also exhibits L/R symmetry [44]. This astounding regularity points to existence of yet undiscovered mechanism that must precisely regulate the phases sequence in the full cycle of changes in the cortical MTs orientation. Must be also independent of the other, hypothetical one, which controls L/R symmetry of cambial cellular events such as oriented anticlinal divisions and intrusive growth. This assumption is supported even further by the S helix of the secondary wall thickenings in differentiating protoxylem. The first, S1 deposit of this wall, not yet completely covering the primary wall surface prevents the cells, exposed to mechanical stress caused by longitudinal expansion of growing shoot, from breaking (**Figure 2**).

The discovery of cell intrinsic chirality resulted from the fundamental question how the laterality of organs within the animal body is accomplished. Over the years much attention has been paid to this phenomenon and its connection with the development of L/R symmetry of the whole multicellular organism. It was found that blood neutrophiles polarity, defined by position of centrosomes with regard to the cell nucleus, makes them capable of directional movements in absence of polar external signals. This property disappears after application of drugs affecting MT function [45].

The model invertebrate organism *Drosophila melanogaster* provided evidence that myosin encoding gene mutation switches cell chirality and results in the development of *situs inversus* phenotype of the hindguts or genitalia [7, 8, 10, 46]. In vertebrates the development of typical L/R asymmetry from anteceding state of embryo bilateral symmetry is generated by various mechanisms. One of them is based on function of axonemal dynein. This motor-protein is responsible for appropriate beating of cilia in nodal epithelial cells, causing the directed ion current. Defects in the gene structure encoding for the dynein results in random selection of heart position in mouse embryo [47]. The involvement of C kinase signaling pathway in the reversal of cells chirality leading to mirrored position of heart was recently shown in chicken embryos [48].

All the above studies show how intrinsic cell polarity may be translated onto the higher level of multicellular organism organization in animals. Very little though is known about L/R symmetry regulation in multicellular plant organisms.

3. Chiral processes and structures in multicellular organisms

The axiality of multicellular organism is similar to the polarity of a single cell but on hierarchically higher level of organization. The first evolutionarily step towards axiality of the integrated biological system composed of many cells represent 1D filaments of cyanobacteria, eukaryotic algae and fungi or moss protonemata. Higher plants maintain this ancestral condition as a kind of atavistic trait at the embryonic stage of their development - linear suspensor, which originates from the basal cell of already polarized and divided zygote, transports nutrients to the 3D globular embryo, developing from the apical cell [18]. Many forms of animals with such model organisms as tiny worm *Coenorhabditis elegans* [49] or fruit fly *Drosophila melanogaster* and finally ourselves, exhibit axiality, metamerism and L/R symmetry.

In this section the short survey of the most interesting cases of mirror symmetry in multicellular plants and animals will be made and the mechanisms that stay behind them will be discussed.

3.1 Changing chirality in the thalli of charophytes

Architecture of these green algae resembles that of the horsetails. The thallus of *Chara* is composed of the giant internodal cells typically enveloped by the sheet of cortical cells, which take an origin from the adjacent nodal cells. From the nodes the 1-st order branchlets grow out horizontally, on which the reproductive structures are positioned: ovaloid oogonia and spherical antheridia. They may be treated as the 2-nd order axial outgrowths of the branching thallus. The peculiar, to date unexplained transition takes place from the Z orientation of enveloping cortical cells in the main axis of the thallus, through their mostly parallel alignment in the branchlets, to the S orientation in oogonia (**Figure 3**). The developmental sequence of these changes in the chiral structure of extant *Chara* species is always the same, although gyrogonites (fossilized oogonia) of charophytes show that in the late Devonian period the right-handed (Z) oogonia were also present. They belonged to the charophyte family Trochiliscaceae and got extinct with the onset of Mesozoic

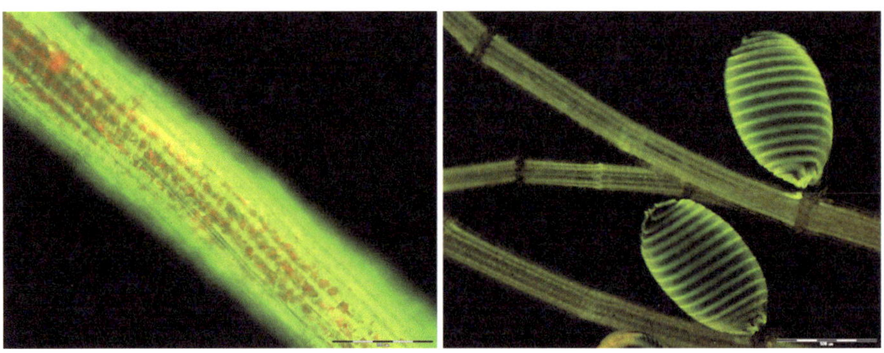

Figure 3.
Details of Chara *sp. thallus architecture. Left photo shows the main axis covered by Z oriented cortical cells; on the right photo two S chiral oogonia sit on the branchlets enveloped by the cortical cells, which run parallel to their axes.*

era [50–52]. We will probably never know if they had the same sequence of chiral changes, but *in reverso*, like the extant *Chara* branching thalli.

The cortical cells of *Chara* parallel orientation of the cytoplasmic streaming in the central internodal cell they envelop. How do they read this direction? How is this information translated from the interface between ecto- and endoplasm, where the cellular engine of cyclosis is located [39, 53, 54], to the surface of the cell wall, on which the enveloping cells slide? Finally, how and why does it change in the *Chara* branching system, from the right in the main axis to the left in oogonia? Is actin – motor protein involved? Actin microfilaments must then have orientation of their alignment determined by the position of the cell in branching thallus. How? These problems remain unresolved.

3.2 Oriented cell divisions in apical cells of mosses and ferns

The control over orientation of the cell division plane is of particular importance in plant tissues. Plant protoplasts are "imprisoned" within the boxes of their cell walls. They cannot migrate as freely as do the animal cells during embryonic stages of ontogenetic development. Morphogenesis of plants relies entirely on the properly oriented cell divisions.

In the simple multicellular filaments of green algae we encounter for the first time the manifestation of L/R symmetry. The plane of cell divisions may be inclined relative to the filament axis either to the left or to the right as exemplified by the filamentous green alga *Coleochaete nitellarum* [55]. Also in planar (2D) gametophytes of ferns the pyramidal apical cell (AC) with rectangular base divides alternately to the left and to the right in a regular sequence (**Figure 4**).

Precision in controlling chiral configuration of cell divisions is even more striking in 3D leafy gametophores of mosses. Their tetrahedral AC, watched from its triangular base, divides either CW or CCW and this direction is randomly established in the development of the gametophore main axis. However, it is not so in the case of its lateral branches – the chirality of their AC is always opposite to that of the supporting axis [5]. It is possible that this antidromic correlation triggers the horizontal gradient of some putative signals vertically transported from the neighboring leaves of gametophore (**Figure 5**). Two genes of the model moss *Physcomitrella patens* were identified to be engaged in a process of cell divisions in leafy gametophore: *PpTONE1* controlling intracellular organization of MT cytoskeleton and *PpNOG1* assuring proper development of AC [56, 57].

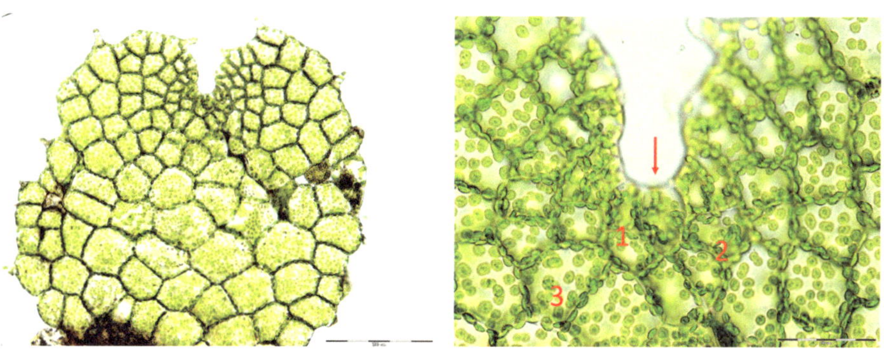

Figure 4.
Typically heart shaped fern gametophyte. This planar structure develops due to activity of AC (red arrow) located atop of its symmetry axis and cleaving derivatives alternately to the left and to the right. On the right photo they are numbered decreasingly.

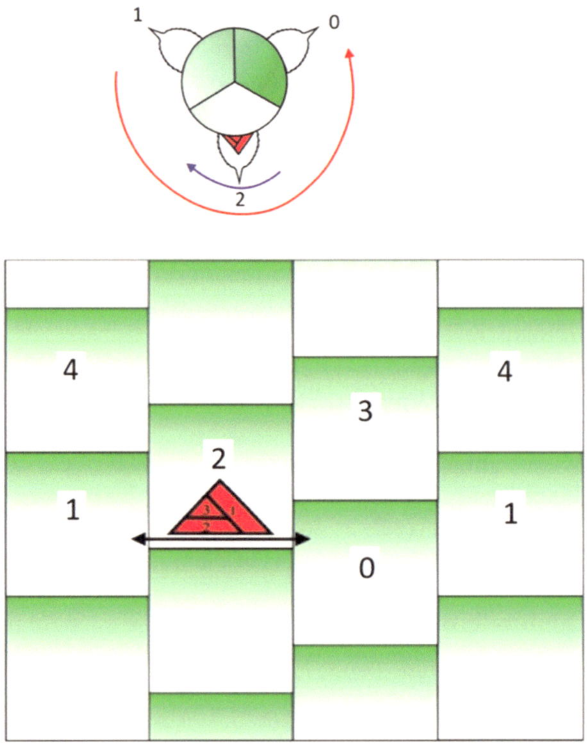

Figure 5.
Scheme of the split open surface of cylindrical moss leafy gametophore. It explains horizontal gradient of the hypothetical signal (green) originating from the leaf segments cleaved CCW (upper red arrow) by the tetrahedral AC of the main axis. The lateral branch AC (red), reading the gradient, starts cleaving its segments CW (upper blue arrow). Main axis segments are numbered increasingly.

3.3 "Music of trees"

Anticlinal, pseudotransverse divisions of elongated stem cells in cambium, cylindrical meristem located in trees between the outer bark and the inner secondary xylem, are chiral. Their partitions, while watched from outside of the tree, are inclined to the right (Z divisions) or to the left (S divisions). Cambium therefore is a plant tissue that exhibits clear L/R symmetry. Also subsequent growth of the cells shortened by the divisions is oriented. Cambial cellular events are not randomly distributed over the surface of the meristem but orderly segregated into domains of opposite chirality. The S and Z domains alternate along the vertical axis of the cambial cylinder [58, 59]. This leads to emergence of structural waviness within which the cells assume alternately the opposite S and Z orientation. Because cambial structure is replicated every year in the annual wood increment, the history of all these developmental changes is recorded in the wood and may be extracted for the periods equaling the age of a tree. The domain pattern and resulting structural waviness are propagated vertically in cambium thus the cells in a particular location undergo the cycle of inclination change (**Figure 6**). This barely known biological rhythm is the longest in nature. Its period approximates 20 years although in some cases may be shorter. Theoretical model predicts that propagation of cell oscillations associated with the domain pattern motion may lead to development of the spiral grain in a tree trunk [14]. Handedness of the spiral grain should depend, according to the model, on the nature of the wave front i.e. the direction of the first change in stem cells inclination during the initial period (**Figure 7**).

Neither the molecular mechanisms, nor the nature of domain positional information for dividing and growing cells have been elucidated so far. The first suspect

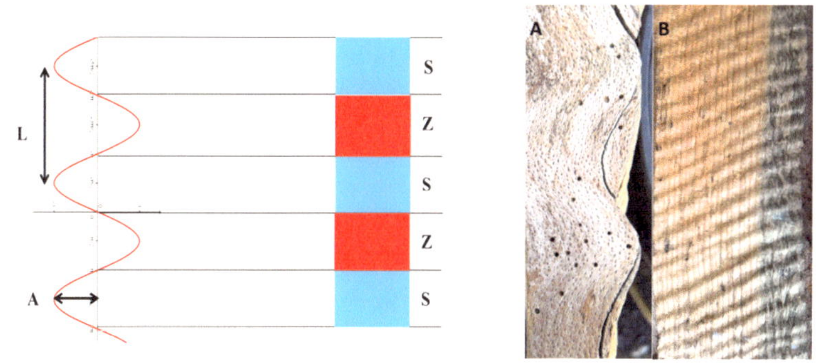

Figure 6.
Scheme of the relationship between hypothetical cambial morphogenetic wave and the domain pattern composed of S (blue) and Z (red) domains. They lead to development of wavy cambium and subsequently to wavy wood. (A) Tangential face of the beech wood, (B) radial split face of the oak wood showing dynamics of the wavy pattern: inclined ripples indicate upward movement of the structural waviness in cambium - bark is on the right side of the photo.

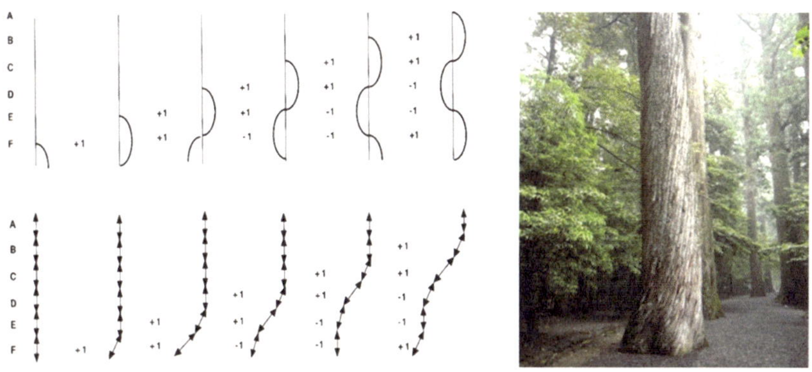

Figure 7.
Scheme explaining how morphogenetic wave propagated upward and exciting oscillations of cambial cells may lead to development of spiral grain in such trees as majestic sugi tree (Cryptomeria japonica) *on the right photo.*

is the polar auxin transport changing directions due to redistribution of the hormone carriers in plasma membrane of cambial stem cells. The intrinsic cell chirality resulting from unknown nature of the intracellular oscillator cannot be excluded. In one tree more than one domain pattern may be present – they usually differ in the domain size and propagation velocity. According to the hypothesis put forward by their discoverer, the domain patterns result from morphogenetic waves traveling in the tissue and capable of superposition [59]. This means that the trees play silent music, the beauty of which is mostly unknown even to the scientists.

3.4 Twinning vines

The helical growth of plant organs is not uncommon [15]. Some plants developed quite effective strategy to grow quickly towards the light source relying on the support, provided sometimes even by another plant. This way they do not have to spend too much energy for building sturdy skeleton composed of mechanical tissues. Finding support is possible thanks to circumnutation of the shoot tip, caused by differential growth along the circumference [60]. Some vines are heterochiral, i.e. capable of twinning CW and CCW. Around 90% of the homochiral species twirl CCW [61, 62] but in some genera the direction of twinning is a species specific trait. *Wisteria sinensis* and Japanese *Wisteria floribunda,* both the attractive ornamental vines, are opposite in this respect. Darwin, who was very interested in twinning plants and made observations on *W. sinensis* stated: "*I have seen no instance of two species of the same genus twinning in opposite directions, and such cases must be rare*" [61]. He could say so because in his times Japanese *Wisteria* has not yet been introduced to England.

The confusing descriptions of *Wisteria* in botanical literature [63] show clearly that there is a certain problem with definition of a chiral configuration of structures or processes with mirror symmetry in biological systems. *W. sinensis* ascending shoot twins CW when looked at from its base and CCW when looked at from above. The same configuration of this plant twining, in some sources is claimed to be CCW [64] in others CW [65]. In Darwin's words the plant "*moves against the sun*" [61]. Compton and Lack [63] claim that *W. floribunda*:" ... has climbing woody stems twining from left to right...", which should not be if it is truly opposite to *W. sinensis*. It all shows the importance of clear convention how the chiral configuration is determined. Moving along the S helix upward is a CW motion whereas descending along the same line we move CCW. While looked at from outside the *W. sinensis* twins from the left to right (Z configuration). It is opposite (S configuration) when looked at from the inside of the growing shoot's helical structure. The same necessity of defining chiral configuration according to specific convention applies to the cellulose microfibrils rotated in the layers of the plant cell secondary wall, to the spiral grain in a tree trunk or to the cells enveloping charophyte oogonia. It seems that definition of the helix chiral configuration, looked at from its outside, as being S or Z is the most reasonable and unequivocal.

Molecular mechanism responsible for the direction of plant climbers twinning is not known. The results of the studies on the *lefty* and *spiral* mutants of the model plant *Arabidopsis thaliana* [40–42] suggest the involvement of the genetic factor. It is possible that the species specific behavior depends on distinct and constitutive gene expression patterns established differently for each species.

3.5 Aestivation

The petal folding in a flower bud, in most of the flowering plants, is clearly chiral. Petals overlap either CCW or CW and this chiral configuration is often later maintained in fully developed flower (**Figure 8**). The direction of petals

Figure 8.
The chiral CW folding (aestivation) of petals in Hawaiian plumeria's flower bud (left) is maintained in a pinwheel-like corolla of an open flower (right).

folding may be, like in the case of circumnutation, the species specific trait. For instance, *Anagallis arvensis* petals always twist CCW, whereas in Hawaiian plumerias they do it otherwise. Common European weed *Malva neglecta* in turn is heterochiral, capable of producing CW and CCW buds on the same individual plant.

3.6 Phyllotaxis

Among best known chiral phenomena and investigated since the ancient times [66] is helical phyllotaxis – the regular distribution of lateral organs such as leaves or flowers on a plant shoot. Their consecutive primordia, circumferentially equidistant, emerge on the vertically growing shoot apex in the regular intervals. The primordia may be connected with an imaginary line called ontogenetic helix. The helix S or Z configuration depends on whether the process of primordia initiation proceeds CW or CCW. The plantlets growing from seeds have this configuration established at random in the main axis. It is not so, however, in the axes of lateral branches. Their ontogenetic helix may be either concordant (a homodromy case) or discordant (an antidromy case) with that of the supporting axis. It has been found, that even when both phyllotactic correlations occur with the same frequency [67] the supporting axis and the laterals may have the same chirality of vascular sympodia - elements of the internal transport system strongly related to phyllotaxis (**Figure 9**).

The sympodia follow the course of one set of superficial secondary helices - phyllotactic prastichies. Two sets of parastichies running in opposite directions constitute a phyllotactic lattice. This is why even when ontogenetic helices in two axes making up one branching unit are discordant, the axes still may be concordant on the level of their vasculature. The numbers of parastichies in the sets of opposite chiral configuration belong to the mathematical series, the quality of which is associated with the size of circumferential distance between successive primordia. This distance, usually given in an angular measure, is known as divergence angle. The most common is the main Fibonacci series (1,1,2,3,5,8,13...) present in the system with the divergence angle approximating 137,5 degrees or Lucas series (1,3,4,7,11 ...) with the angle close to 99,5 degrees. There are also many other divergencies and phyllotactic patterns [68].

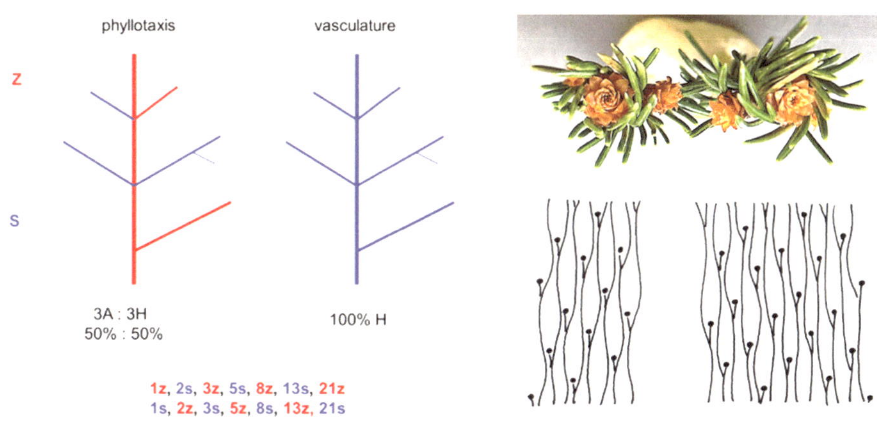

Figure 9.
Scheme on the left shows how, in the laterals of one coniferous branching shoot, ontogenetic helix may be either S (blue) or Z (red) but orientation of vascular sympodia the same in the whole system. The sympodia chiral configuration depends on their number, which is one of the mathematical series shown below the scheme. H- homodromic, A - antidromic correlations of chiral configurations. Upper right photo shows the righthanded and lefthanded whirls of needles in two coniferous shoots with the same S Fibonacci phyllotaxis. Their opposite chiral configurations, resulting likely from growing shoot rotation, are caused by the different sympodia numbers and orientations shown below.

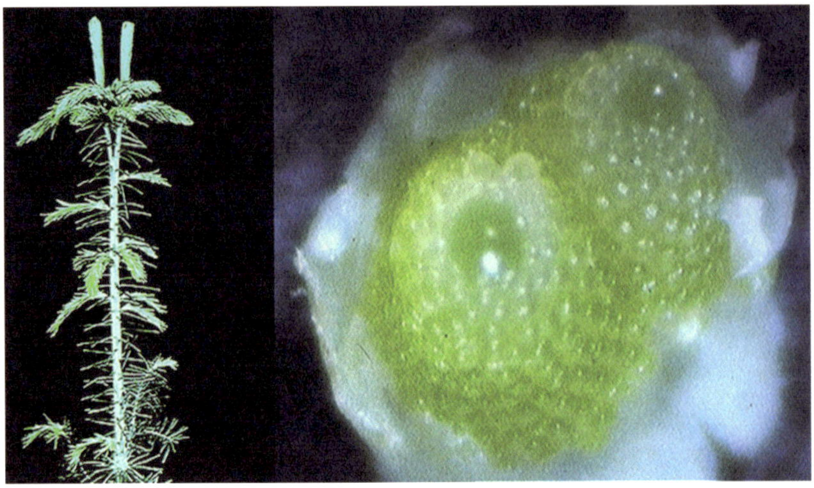

Figure 10.
*Shoot apical meristems isolated from winter buds of balsam fir (*Abies balsamea*) are truly green living crystals. The needle primordia are tightly packed on their lateral surface resembling crystal lattice. The unique case shown here illustrates the atavistic trait of dichotomy, rare in otherwise strictly monopodial conifer.*

Asymmetry of phyllotactic lattice with regard to the shoot axis is most probably responsible for the peculiar twirling of needles frequently seen on the top of coniferous shoot (**Figure 9**). The chirality of these twirls results likely from the growing shoot torsion and is rather related to the orientation of vascular sympodia than to the chiral configuration of ontogenetic helix.

Regularity of primordia initiation resembles crystal growth. The plant apical meristem where the primordia are tightly packed may be called by *licentia poetica*, a living crystal [69] (**Figure 10**). The similarity has been strengthened by the discovery that in phyllotactic lattices dislocations occur [68–70]. Single dislocation often changes not only a quality of the pattern but, most importantly, the chirality of ontogenetic helix (**Figure 11**).

3.7 Snail shell, narwhal tooth and ourselves

Chirality of spiral snail shells has intrigued the scientists for centuries not less than the regularity of phyllotaxis. One of the memorable episodes from the Jules Verne's

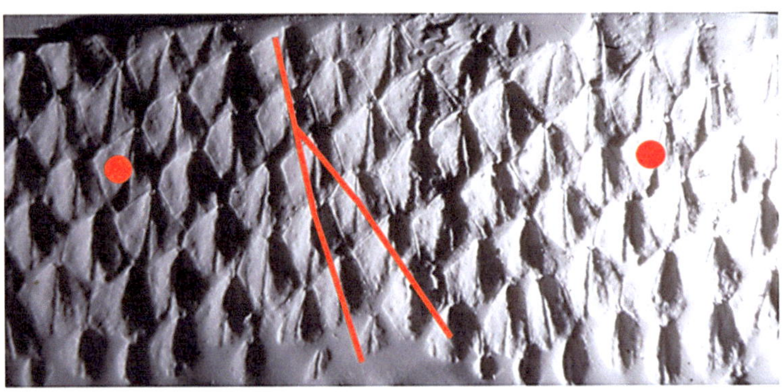

Figure 11.
Modeling clay replica of magnolia's reproductive shoot shows single dislocation (red lines) in the phyllotactic lattice. This developmental event changes here not only the phyllotactic pattern but also the chiral configuration of ontogenetic helix. Red dots label the same pattern element replicated twice on both sides of the unrolled surface of the shoot; it enables counting the numbers of parastichies in two opposite sets; the numbers change from 5:9 to 5:8.

famous novel *Twenty Thousand Leagues Under the Seas* tells the story of Professor Aronnaux finding the extreme rarity – lefthanded shell of the olive snail. It is known to malacologists that approximately 90% of all gastropods have their shells righthanded – of Z type. However, there are snails like *Amhidromus inversus* that have dextral and sinistral shells equally frequent, or like *Neptunea angulata* where the shells are exclusively sinistral. Notably the Z shell grows as the descending spiral, coiling from the top downward. Therefore moving downward the dextral spiral of the shell we execute CW motion not CCW as it would be in the case of ascending helices of plant structures.

The direction of the shell coiling is initiated in the embryo by the spiral cell cleavage typical for lower Metazoa including snails. At this stage there is a possibility of altering the normal pattern and forcing experimentally the development of opposite chirality. The genetic mechanisms determining the chirality patterning in snails are slowly being unraveled through studying specific gene expression patterns in wild type organisms and mutants [71, 72].

However, the reason for a change in a frequency of shell chiral configuration among individuals within a population sometimes can be truly surprising. It was found that among small, properly righthanded *Satsuma* snails the opposite, lefthanded individuals started growing in number. Thorough studies revealed that it was due to activity of predators [73]. The snakes *(Pareas iwasakii)* with their asymmetric jaws, preferably eating the righthanded snails, decimated their population. Through the selective elimination of these snails from the initial population, the snakes contributed to the prevalence of rare lefthanded snails. The divergence of the initial population led to development of a new species. What a wonderful example of Darwinian selection!

In contrast to the snails equipped with molecular mechanism allowing for development of S or Z shells another famous chiral structure in animal kingdom, the narwhal tusk exhibits always a lefthanded spiral (S). It is so even in a case of both tusks being fully developed in one individual, which is rare. The opposite spiral is perhaps also possible as shows the walking stick used by Darwin displayed in the collections of Science Museum, UK [74]. Either the material it is made of is not a narwhal tusk or Darwin, who was very interested in cases of mirror-symmetry in nature [61], consciously adopted and used this particular object being aware of its uniqueness. The third, least probable possibility is that the artist carved the right spiral from the polished left spiral of the narwhal tusk. The reason for the prevalence of one chiral configuration in spiraling of the narwhal tusks is unknown.

L/R symmetry of our body is best illustrated by asymmetry of the internal organ positions like heart or liver. It is our hands, however, that are most frequently presented as an example of mirror symmetry. Less known aspect of the symmetry in a human body is the hair whorl resembling, with all due proportion, the twirling needles in coniferous shoot (**Figure 9**). There is a dispute over significance of the observation that the righthanded people have more frequently their hair twirling CW on the top of their heads. Half of the lefthanded people have in turn the CCW hair whorls. Until now the search for possible common, genetic etiology of these chiral phenomena has been unsuccessful [75]. There is also unknown whether the scalp hair whorl chirality is concordant or discordant with the smaller whorls of minute hair covering our whole body. These become visible, especially in children, when their skin is suntanned.

4. Conclusion

There is no one universal mechanism that stays behind the mirror symmetry of life. The frequency of both chiral configurations is not the same in different biological systems. However, as it has been discussed here, it was not always assessed carefully enough by investigators. *Spirogyra* case is uncertain. The narwhal tusks are probably always S-helical. On the contrary the shell coiling in most of the snails is of Z type. In many systems the chiral configuration of structures or processes is strictly controlled, in many others the control is loose or absent, which results in equal frequency of both S and Z forms.

In light of the newly discovered intrinsic cell chirality in animal cells we have now a great perspective of disentangling the ultracellular and molecular basis for the dynamic wavy and spiral patterns developing in cambia of trees - one of the most intriguing and least known biological rhythms. Discovered and thoroughly characterized by Hejnowicz and his followers in the last decades of past century it still remains a great mystery. Neither the nature of specific positional signals coming from the dynamic morphogenetic field to the cambial stem cells nor the mechanism of their response i.e. S or Z oriented cell divisions on the cylindrical surface of this embryonic tissue, have been elucidated. The tissue is also intriguing because of its structure. It may be compared to that of the liquid crystals – the elongated cambial stem cells may be aligned in the regular horizontal tiers, like the molecules in the smectic phase of the liquid crystal, or irregularly but parallel to the vertical axis, like in the nematic phase. The oscillating cambial cells taken together with their derivatives, continuously rotated in the successive wood layers, resemble the third, cholesteric phase of the liquid crystal [76].

Biomechanics of structures based on the possibility of changing chiral configurations, clearly the adaptive trait, cannot be underestimated. The resulting interlocked systems provide high resistance to mechanical stress. Interlocked are the cellulose microfibrils in the successive layers of the secondary cell wall in a single cell and, on the macroscale, the oppositely oriented wood fibers in the packets of consecutive wood increments of such giants as mahogany or camphor trees. Also the system of cortical resin canals in the young coniferous shoots, which runs oppositely to oriented vascular sympodia strengthens the axis mechanically.

These examples together with the crystalline character of phyllotactic patterns and cambial cells arrangements bring us back to already mentioned, at the beginning of this chapter, universality of some solutions based last but not least on the presence of mirror symmetry of life.

Acknowledgements

The author expresses sincere thanks to her long life associates in the Department of Plant Developmental Biology at the University of Wroclaw, Poland for their continuous inspiration and support; especially to Dr. Katarzyna Sokołowska, who helped here with **Figure 5**, Dr. Alicja Banasiak, expert on plant cell walls and polar auxin transport and to Magdalena Turzańska (MSc), an excellent bryologist, for our endless discussions on the hidden beauty of a small world she immortalizes on her artistic microphotographs.

Author details

Beata Zagórska-Marek
University of Wrocław, Wrocław, Poland

*Address all correspondence to: beata.zagorska-marek@uwr.edu.pl

References

[1] Halebian M, Morris K, Smith C. Structure and Assembly of Clathrin Cages. Sub-cellular biochemistry. 2017;83:551-567. DOI:10.1007/978-3-319-46503-6_20

[2] Debut A, Guerra S, Kleber A. Fullerene-Based Symmetry in *Hibiscus rosa-sinensis* Pollen. PLoS ONE. 2014; 9: e102123. DOI:10.1371/journal.pone.e102123

[3] Zagórska-Marek B. Plant Meristems and Their Patterns. In: Carbone A, Gromov M, Prusinkiewicz P, editors. Pattern formation in biology, vision and dynamics. Singapore, New Jersey London, Hong Kong: World Scientific; 2000. p. 217-239. DOI: 10.1142/9789812817723_0011

[4] Canejo J, Godinho M. Cellulose Perversions. Materials (Basel, Switzerland). 2013;6(4):1377-1390. DOI: 10.3390/ma6041377

[5] Zagórska-Marek B, Sokołowska K, Turzańska M. 2018. Chiral events in developing gametophores of *Physcomitrella patens* and other moss species are driven by an unknown, universal direction- sensing mechanism. American Journal of Botany.2018;105(12): 1-9. DOI:10.1002/ajb2.1200

[6] Gilbert SF. Early Development of Snails. In: Developmental Biology. 6th edition. Sunderland (MA): Sinauer Associates; 2000. Available from: https://www.ncbi.nlm.nih.gov/books/NBK10074/ [Accessed: 2021-01-30]

[7] Inaki M, Liu J, Matsuno K.. Cell chirality: its origin and roles in left-right asymmetric development. Philos. Trans. R. Soc. Lond. B Biol. Sci. 2016;371:20150403. DOI: 10.1098/rstb.2015.0403

[8] Inaki M, Sasamura T, Matsuno K. Cell Chirality Drives Left-Right Asymmetric Morphogenesis. Front. Cell Dev. Biol. 2018;6:34. DOI: 10.3389/fcell.2018.00034

[9] Fan J, Zhang H, Rahman T, *et al.* Cell organelle-based analysis of cell chirality, Communicative & Integrative Biology.2019; 12(1):78-81. DOI: 10.1080/19420889.2019.1605277

[10] Ishibashi T, Inaki M, Matsuno K. Statistical Validation Verifies That Enantiomorphic States of Chiral Cells Are Determinant Dictating the Left- or Right-Handed Direction of the Hindgut Rotation in *Drosophila*. *Symmetry.* 2020; *12(12)*:1991. DOI: 10.3390/sym12121991

[11] Hejnowicz Z. Morphogenetic waves in cambia of trees. Plant Sci. Lett. 1973;1: 359-366. DOI: 10.1016/0304-4211(73)90060-6

[12] Hejnowicz Z, Romberger J. Migrating cambial domains and the origin of wavy grain in xylem of broadleaved trees. American Journal of Botany. 1973;60:209-222. DOI: 10.1002/j.1537-2197.1973.tb10218.x

[13] Harris J. Spiral Grain and Wave Phenomena in Wood Formation. Berlin. Heidelberg: Springer-Verlag; 1989. 215p. DOI:/10.1007/978-3-642-73779-4

[14] Zagórska-Marek B. Morphogenetic waves in cambium and figured wood formation. In: Iqbal M, editor. Encyclopedia of plant anatomy: The cambial derivatives. Berlin. Stuttgart: Gebrüder Borntraeger;1995. p. 69-92.

[15] Smyth D. Helical growth in plant organs: mechanisms and significance. Development. 2016;143:3272-3282. DOI: 10.1242/dev.134064

[16] Shaw SL, Quatrano RS. The role of targeted secretion in the establishment of cell polarity and the orientation of

the division plane in *Fucus* zygotes. Development. 1996;122(9):2623-2630.

[17] Hable W, Miller N, Kropf D. Polarity establishment requires dynamic actin in fucoid zygotes. Protoplasma. 2003; 221:193-204. DOI: 10.1007/s00709-002-0081-0

[18] Souter M, Lindsey K. Polarity and signaling in plant embryogenesis. *Journal of Experimental Botany.* 2000;51(3470):971-983. DOI:10.1093/jexbot/51.347.971

[19] Gonzales-Reyes A, Elliott H, St. Johnston D. Oocyte determination and the origin of polarity in *Drosophila*: the role of the spindle genes. Development. 1997;124(24):4927-4937

[20] Hosseini SM, Moulavi F, TanhaieVash N, *et al.* Evidence of Oocyte Polarity in Bovine; Implications for Intracytoplasmic Sperm Injection and Somatic Cell Nuclear Transfer. Cell Journal. 2017;19(3):482-491. DOI:10.22074/cellj.2017.4887

[21] Takata K. Glucose Transporters in the Transepithelial Transport of Glucose, *Journal of Electron Microscopy*. 1996;45(4):275-284. DOI: 10.1093/oxfordjournals.jmicro.a023443

[22] Berkel K, Boer R, Scheres B, Tusscher K. Polar auxin transport: Models and mechanisms. Development. 2013;140:2253-2268. DOI: 10.1242/dev.079111

[23] Friml J, Vieten A, Sauer M, *et al*. Efflux-dependent auxin gradients establish the apical-basal axis of Arabidopsis. Nature. 2003;426:147-*153.*

[24] Abas L, Benjamins R, Malenica N, *et al*. Intracellular trafficking and proteolysis of the *Arabidopsis* auxin-efflux facilitator PIN2 are involved in root gravitropism. Nat Cell Biol. 2006;8(3):249-256. DOI: 10.1038/ncb1369

[25] Spirogyra Cell [Internet]. Available from: https://sciencemythos.weebly.com/spirogyra-cell.html; [Accessed: 2021-01-11]

[26] Science Photo Library. Spirogyra algae, light micrograph[Internet]. Available from: https://www.sciencephoto.com/media/419721/view; [Accessed: 2021-01-18]

[27] FineArtAmerica. Dennis Kunkel Microscopy/science Photo Library. [Internet]. Available from: https://fineartamerica.com/featured/2-paramecium-multimicronucleatum-dennis-kunkel-microscopyscience-photo-library.html [Accessed: 2021-01-24]

[28] Bell AJ, Satir P, Grimes GW. Mirror-imaged doublets of *Tetmemena pustulata*: implications for the development of left–right asymmetry. Dev. Biol. 2008;314:150-160. DOI:10.1016/j.ydbio.2007.11.020

[29] Tassin AM, Lemullois M, Aubusson-Fleury A. *Paramecium tetraurelia* basal body structure. Cilia. 2016;5: article 6. DOI:10.1186/s13630-016-0026-4

[30] Mattox K, Stewart K. Classification of the green algae: a concept based on comparative cytology. In: Irvine D, John D. editors. Systematics of the Green Algae. London: Academic Press; 1984. p. 29-72.

[31] Miyaji K, Hori T. The ultrastructure of *Spongomorpha duriuscula* (Acrosiphoniales, Chlorophyta), with special reference to the flagellar apparatus. Jap. J. Phycol. 1984;32:307-318

[32] O'Kelly C, Floyd G. The flagellar apparatus of *Entocladia viridis* motile cells, and the taxonomic position of the resurrected family Ulvellaceae (Ulvales, Chlorophyta). J. Phycol. 1983;19:153-164

[33] Van den Hoek C, Mann D, Jahns H. Algae. An introduction to phycology. Cambridge University Press; 1995. 627 p.

[34] Ishijima S, Hamaguchi Y. Calcium ion regulation of chirality of beating flagellum of reactivated sea urchin spermatozoa. Biophysical Journal. 1993;65 (4):1445-1448. DOI: 10.1016/S0006-3495(93)81210-4

[35] Dexter JP, Prabakaran S, Gunawardena J. Complex Hierarchy of Avoidance Behaviors in a Single-Cell Eukaryote. Current biology. 2019;29(24): 4323-4329.E2. DOI: 10.1016/j.cub.2019.10.059

[36] Kuźnicki L, Sikora J. Inversion of spiralling of *Paramecium aurelia* after homologous antiserum treatment. Acta Protozool. 1966;4:263-268

[37] Wan L, Ronaldson K, Park M, *et al*. Micropatterned mammalian cells exhibit phenotype-specific left-right asymmetry. Proc. Natl Acad. Sci. 2011;108:12295-12300. DOI: 10.1073/pnas.1103834108

[38] Shi H, Quint D, Grason G. *et al.* Chiral twisting in a bacterial cytoskeletal polymer affects filament size and orientation. Nature Communications. 2020;11: 1408. DOI: 10.1038/s41467-020-14752-9

[39] Higashi-Fujime S, Ishikawa R, Iwasawa H, *et al*. The fastest actin-based motor protein from the green-algae, *Chara*, and its distinct mode of interaction with actin. FEBS Letters. 1995;375:151-154. DOI: 10.1016/0014-5793(95)01208-v.

[40] Furutani I, Watanabe Y, Prieto R, *et al*. The SPIRAL genes are required for directional control of cell elongation in *Arabidopsis thaliana*. Development. 2000;127:4443-4453.

[41] Hashimoto T. Molecular genetic analysis of left-right handedness in plants. Philosophical transactions of the Royal Society of London. Series B, Biological sciences. 2002;357(1422):799-808. DOI: 10.1098/rstb.2002.1088

[42] Ishida T, Kaneko Y, Iwano M, *et al*. Helical microtubule arrays in a collection of twisting tubulin mutants of *Arabidopsis thaliana*. Proc. Natl Acad. Sci. 2007;104 (20):8544-8549.

[43] Abe H, Funada R, Imaizumi H, *et al*. Ohtani J, Fukazawa K. Dynamic changes in the arrangement of cortical microtubules in conifer tracheids during differentiation. Planta. 1995;197: 418-421. DOI:10.1007/BF00202666

[44] Zagórska-Marek B. Microfibrils orientation in figured wood. In: Proceedings of Pacific Regional Wood Anatomy Conference organized by the Int Assoc of Wood Anatomists and S5.01 of the International Union of Forest Research Organization; 1-7 October 1984; Tsukuba Ibaraki, Japan. Syoji Sudo editor;1984. p.106-108

[45] Xu J, Van Keymeulen A, Wakida N, *et al*. Polarity reveals intrinsic cell chirality. Proc Natl Acad Sci. 2007;104(22):9296-9300. DOI: 10.1073/pnas.0703153104

[46] Ishibashi T, Hatori R, Maeda R, *et al*. E and ID proteins regulate cell chirality and left-right asymmetric development in *Drosophila*. Genes Cells. 2019;24(3):214-230. DOI: 10.1111/gtc.12669

[47] McGrath J, Somlo S, Makova S, *et al*. Two Populations of Node Monocilia Initiate Left-Right Asymmetry in the Mouse. Cell. 2003;114(1):61-73. DOI: 10.1016/S0092-8674(03)00511-7

[48] Poulomi R, Chin A, Worley K, *et al*. Intrinsic cellular chirality regulates left–right symmetry breaking during cardiac looping. Proc Natl Acad Sci. 2018;115

(50): E11568-E11577. DOI: 10.1073/pnas.1808052115

[49] Wood W, Kershaw D. Handed asymmetry, handedness reversal and mechanisms of cell fate determination in nematode embryos. Ciba Found Symp. 1991;162:143-59; discussion 159-64. DOI: 10.1002/9780470514160.ch9

[50] Peck, R. Fossil Charophyta. The American Midland Naturalist. 1946;36(2): 275-278. DOI:10.2307/2421503

[51] Peck R, Morales G. The Devonian and Lower Mississippian Charophytes of North America. Micropaleontology.1966;12(3):303-324. DOI:10.2307/1484549

[52] Soulié-Märsche I. Chirality in Charophytes: Stability and Evolution from 400 million Years to Present. In: Zucchi C, Caglioti L, Pályi G, editors. Advances in BioChirality. 1st ed. Amsterdam, Lausanne, New York, Oxford, Shannon, Singapore, Tokyo: Elsevier Science; 1999. p. 191-207. DOI: 10.1016/B978-008043404-9/50013-9

[53] Kamiya N, Kuroda K. Velocity distribution of the protoplasmic streaming in *Nitella* cells. Bot Mag Tokyo. 1956;69:544-554.

[54] Tominaga M, Ito K. The molecular mechanism and physiological role of cytoplasmic streaming. Curr Opin Plant Biol. 2015;27:104-110. DOI: 10.1016/j.pbi.2015.06.017.

[55] *Coleochaete nitellarum* [Internet]. Available from: http://science.umd.edu/labs/delwiche/Strp/Chlorophyta/charophyceae/Cnit.jpg [Accessed: 2021-01-26]

[56] Spinner L, Pastuglia M, Belcram K, *et al*. The function of TONNEAU1 in moss reveals ancient mechanisms of division plane specification and cell elongation in land plants. Development. 2010;137:2733-2742. DOI: 10.1242/dev.043810

[57] Moody L, Kelly S, Rabbinowitsch E, *et al*.. Genetic regulation of the 2D to 3D growth transition in the moss *Physcomitrella patens*. Current Biology. 2018;28:473-478. DOI: 10.1016/j.cub.2017.12.052

[58] Hejnowicz Z. Upward movement of the domain pattern in the cambium producing wavy grain in *Picea excelsa*. Acta Societatis Botanicorum Poloniae. 1971;40(3):499-512. DOI: 10.5586/asbp.1971.037

[59] Hejnowicz, Z. Pulsation of domain length as support for the hypothesis of morphogenetic waves in the cambium. Acta Societatis Botanicorum Poloniae, 1974;43(2):261-271. DOI: 10.5586/asbp.1974.025

[60] Goriely A, Neukirch S. Mechanics of Climbing and Attachment in Twining Plants. Phys. Rev. Lett. 2006;97: article 184302. DOI: 10.1103/PhysRevLett.97.184302

[61] Darwin C. On the movements and habitats of climbing plants. J. Linnean Soc. (Botany). 1867;9:1-118.

[62] Edwards W, Moles AT, Franks P. The global trend in plant twining direction. Glob Ecol Biogeogr. 2007;16:795-800.

[63] Compton JA, Lack H W.The discovery, naming and typification of *Wisteria floribunda* and *W. brachybotrys (Fabaceae)* with notes on associated names. – Willdenowia. 2012;42:219-240. 10.3372/wi.42.42207. Available from: https://bioone.org/journals/Willdenowia [Accessed: 2021-01-27]

[64] *Valder P. Wisterias: a comprehensive guide. Portland, Or.: Timber Press. 1995.* ISBN 0881923184. OCLC 32647814

[65] Haldeman J. As the wine twines. Native and Naturalized Plants of the

Carolinas and Georgia [Internet]. 2007. Available from http:// www.namethatplant.net/article_ asthevinetwines.shtml [Accessed: 2021-01-04]

[66] Adler I, Barabe D, Jean RV. A History of the Study of Phyllotaxis. *Annals of Botany*. 1997;80(3): 231-244. DOI: 10.1006/anbo.1997.0422

[67] Banasiak A, Zagórska-Marek B. Structural Integrity of Vascular System in Branching Units of Coniferous Shoot. Acta Societatis Botanicorum Poloniae. 2020;89(1): article 8915. DOI: 10.5586/ asbp.8915

[68] Zagórska-Marek B. Phyllotaxic diversity in Magnolia flowers. Acta Societatis Botanicorum Poloniae. 1994;63:117-137. DOI: 10.5586/ asbp.1994.017

[69] Zagórska-Marek B. Magnolia flower – the living crystal [Internet]. Magnolia. 2011;89:11-21. Available from: https:// sophia.smith.edu/blog/phyllotaxis/ files/2014/07/Magnolia-Society-I-ISSUE-89-BEATA.pdf [Accessed: 2021-01-30]

[70] Zagórska-Marek B, Szpak M. The significance of γ-and λ-dislocations in transient states of phyllotaxis: how to get more from less – sometimes! Acta Societatis Botanicorum Poloniae. 2016;85(4):article 3532. DOI: 10.5586/ asbp.3532

[71] Noda T, Satoh N, Asami T. Heterochirality results from reduction of maternal *diaph* expression in a terrestrial pulmonate snail. Zoological Letters. 2019;5(1): 2. DOI: 10.1186/ s40851-018-0120-0

[72] Abe M, Kuroda R. The development of CRISPR for a mollusc establishes the formin Lsdia1 as the long-sought gene for snail dextral/sinistral coiling. Development 2019; 146:article dev175976. DOI: 10.1242/dev.175976

[73] Hoso M, Kameda Y, Wu SP, Asami T, Kato M, Hori M. A speciation gene for left–right reversal in snails results in anti-predator adaptation. Nature Communications. 2010;1:article133. DOI: 10.1038/ncomms1133

[74] Science Museum Group Collection. Walking stick made from narwhal tusk with carved ivory pommel, once owned by Charles Darwin, probably English, 1839-1881 [Internet]. Available from: https://collection.sciencemuseumgroup. org.uk/objects/co126728/walking-stick-made-from-narwhal-tusk-with-carved-i-walking-sticks [Accessed: 2021-01-19]

[75] Perelle IB, Ehrman L, Chanza M. Human handedness and scalp hair whorl direction: no evidence for a common cause. Laterality. 2009;14(1):95-101. DOI: 10.1080/13576500802387692.

[76] Oxtoby et al. 23.3: Liquid Crystals [Internet].2020. Available from: https://chem.libretexts.org/ Bookshelves/General_Chemistry/ Map%3A_Principles_of_Modern_ Chemistry_(Oxtoby_et_al.)/ UNIT_6%3A_MATERIALS/23%3A_ Polymeric_Materials_and_Soft_ Condensed_Matter/23.3%3A_Liquid_ Crystals [Accessed: 2021-01-28]